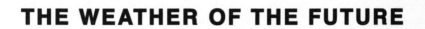

# THE WEATHER OF THE FUTURE

THE WEATHER OF THE FUTURE

# THE WEATHER OF THE FUTURE

## Heat Waves, Extreme Storms, and Other Scenes from a Climate-Changed Planet

## Heidi Cullen

HARPER

*An Imprint of* HarperCollins*Publishers*
www.harpercollins.com

HarperCollins books may be purchased for educational, business, or sales promotional use. For information, please write: Special Markets Department, HarperCollins Publishers, 10 East 53rd Street, New York, NY 10022.

The cover represents an artist's rendering of New York City with higher sea levels. Sea level rise of this magnitude will likely occur several centuries from now as the West Antarctic and Greenland ice sheets melt and higher temperatures warm and expand our oceans.

Special thanks to Lisa Ammerman for her skilled hand in making the maps throughout this book.

FIRST EDITION

*Designed by Joseph Rutt*

Library of Congress Cataloging-in-Publication Data is available upon request.

ISBN 978-0-06-172688-0

10 11 12 13 14    ID/RRD    10 9 8 7 6 5 4 3 2 1

For my fellow scientists.
Never stop seeking the truth.

The sweetest thing in all my life has been the longing to find the place where all the beauty came from.

—C. S. Lewis

# CONTENTS

# LIST OF MAPS

# INTRODUCTION

*"What's the forecast?"*

I heard this question a lot when I first started at The Weather Channel in 2003. People figured that if I worked at a 24-7 weather network, I must be a meteorologist. That was fine by me, I've always been a closet weather geek, and besides, who doesn't secretly want to be a meteorologist? There is something so appealing about going to work every day and predicting the future without the use of tarot cards or constellations.

I'm not sure how many people secretly want to be climatologists, but that's what I really am. And for anyone wondering what a climatologist is, here's a rough answer: Climatologists pick up where meteorologists leave off. We focus on weather timescales beyond the memory of the atmosphere, which is only about one week. And I guess you could say we also focus on timescales beyond human memory, which is shorter than you might think.

Climatologists look at patterns that range from months to hundreds, thousands, and even millions of years. The single most important and most obvious example of climate is the seasonal cycle, otherwise known as the four seasons. The seasons, a result of the 23.5° tilt of the Earth's axis, affect the weather dramatically. And the physics behind the seasons are well nailed down. Summer, the result of one hemisphere's being tilted closer to the sun, is warmer. And winter, the result of the other hemisphere's being tilted away from the sun, is colder. The forecast for the seasons follows the physics; this is why, if I issue a forecast in January that says it will be significantly warmer in six months, you probably won't think I'm a genius, but you'll believe me.

There are countless other patterns on our planet that influence the weather. Take El Niño, for example. Nicknamed by fishermen along the coast of Peru after the Christ child, El Niño is a warm ocean current that typically appears every few years around Christmastime and lasts for several months. From its home in the tropical Pacific Ocean, El Niño is powerful enough to influence weather across the entire world. Unlike the seasons, which are controlled by astronomical forces, El Niño results from processes that happen here on Earth and that don't come and go like celestial clockwork. Climatologists have come to understand that the physics of El Niño result from a series of complex interactions between the ocean and the atmosphere. Technically, El Niño (EN) describes the ocean component, whereas the atmospheric component is known as the Southern Oscillation (SO). That's why climatologists generally refer to it as ENSO.

During an ENSO event, the easterly trade winds weaken and the surface water off the coast of Peru and Ecuador warms up several degrees. This warmer water leads to increased evaporation, causing the air above it to rise and thereby affecting the winds. This conversation between ocean and atmosphere is nuanced and far-reaching. The atmosphere feels the influence of the warm ocean surface below it and conveys the message by shifts in tropical rainfall, which in turn affect wind patterns over much of the globe. For example, most El Niño winters are milder over western Canada and parts of the northern United States, and wetter over the southern United States from Texas to Florida. In other parts of the world, ENSO can bring drought to northern Australia, Indonesia, the Philippines, southeastern Africa, and northern Brazil. Heavier rainfall is often seen in coastal Ecuador, northwestern Peru, southern Brazil, central Argentina, and equatorial eastern Africa.

El Niño is just one of the ways climate can work itself into the weather. You might say meteorologists are obsessed with the atmosphere, whereas climatologists are obsessed with everything that *influences* the atmosphere. But in the end, we're all obsessed with this notion of predicting the future. The atmosphere may be where the weather lives, but it speaks to the ocean, the land, and sea ice on a regular basis. Consider them influential friends that are capable of forcing the atmosphere to behave in ways that are sometimes, as in the case of ENSO, predictable. The hope is that if scientists can untangle all the messy relationships at work within our climate system, we should be better able to keep people out of harm's way. The farther we can extend human memory, the longer out in time a society can see, and the better prepared we'll be for what's in the pipeline.

And that is where global warming enters the picture. If the four seasons are Mother Nature's most prominent signature within the climate system, then you might say that *global warming*, the term that refers to Earth's increasing temperature due to a buildup of greenhouse gases in the atmosphere, is humanity's most prominent signature. The big difference between global warming and other climate and weather phenomena is that in this case, we're the ones doing the talking. And greenhouse gases are the chatter we use to influence the behavior of the atmosphere.

Decades of study suggest that this conversation will slowly drown out all others, its influence cutting across all timescales and all regions of the planet. Global warming has already begun pushing around the timing of the four seasons, and ongoing research shows that it is also influencing the weather. Meanwhile, climate scientists have developed a robust understanding of the physics of this human interaction with the atmosphere. They have collected data and built predictive models of the climate system that are capable of looking into the past and—more important—the future. The forecast for Earth is in, and it's not good.

Part I of this book explains the cutting-edge science behind this long-term climate forecasting, demonstrating why the predictive models for next century should be trusted in the same way that you trust the forecast for tomorrow on your local news. Here we'll examine the relationship between our weather today and our forecasts down the road, looking at how climatologists assess the changing statistics of extreme weather events and how these changing statistics play into the long-term forecast.

We'll also look at the history of weather prediction and how it serves as the foundation of climate forecasts today. Weather forecasts and climate forecasts are based on the same principles of mathematics and physics. Yet they have inherent differences that allow weather forecasts to focus on short-term changes in the atmosphere, whereas climate forecasts focus on long-term changes to the entire system of ocean, land, and ice. Keep in mind: just as my initial forecast that July would be hotter than January didn't involve the weather on a specific day, so too a climate forecast for 2050 or 2100 looks at the big picture and not a specific day. A forecast for the year 2050 has the potential to be as meaningful and as useful as tomorrow's weather forecast; it's just used in a different way.

All this is important because if you don't trust the models, you won't believe the forecasts—and the forecasts are what Part II of this book is all about. In Part II, we'll look at the forty-year forecasts for a few important places around the world. To start, I asked climate scientists to list the places they thought were most vulnerable to the threat of global warming.

I then narrowed the list down to seven key examples. (I've included a fuller list of hot spots identified by climate scientists in Appendix 3.) I chose these seven places not necessarily because they're the most endangered places or because the stories they offer are the most dramatic, but instead because collectively they demonstrate a spectrum of risks that exist with climate change. By mid-century, not every part of the world will be affected by global warming in the same way. Each location I've chosen has its own Achilles' heel, a vulnerability that unabated climate change will expose and exploit until the place is forever altered. Taken together these vulnerabilities show the breadth of repercussions that climate change will bring. It is my hope that whether taken as individual stories or as a whole, the predictions found in this book will demonstrate that global warming will hit all of us in the places we love and the homes where we live.

———

If Hurricane Katrina taught us anything, it is that the worst-case scenario can happen. For the first time in human history, science has given us the ability to peer into a crystal ball of numbers and models and see what kind of a climate we'll be living in by mid-century if we continue to emit carbon at our current levels. I share this look at the future with people outside the scientific community not as a scare tactic or as hyperbole, but because only through such sharing will the world come to understand precisely what is at stake.

Let me show you what I mean. For several years, I've been giving lectures and seminars about climate change to a variety of groups all over the world. Sometimes I speak to scientists; sometimes the audience is primarily students and their parents; sometimes it's politicians and business executives. After one of my first seminars, several years ago, I was standing at the front of the lecture hall, putting my computer away, when I was approached by a man—probably in his late forties—who had a question. He had enjoyed my lecture and found that it opened his eyes to several new aspects of the science and impacts of climate change, but what he really wanted to know was this:

"Do you think I should sell my beach house?"

After he said this, I thought for a moment about how best to calculate the risk associated with owning beachfront property in the United States, factoring in our best estimates of impacts such as sea level rise, storm surge, and saltwater intrusion—just to name a few. The man waited patiently with an earnest, if slightly bemused, look on his face. I suspect he may have been joking with me, but I felt I owed him an answer.

It was then that something hit me: *This is the only way a lot of people can truly connect to the issue of climate change—via a long-term investment like real estate.*

The more I thought about his question, the more I realized that the scientific community had failed to communicate the threat of climate change in a way that made it real for people right now. We, as scientists, hadn't given people the proper tools to see that the impacts of climate change are visible right now and that they go far beyond melting ice caps.

I can honestly say that real estate is what comes up most often when I talk to people about global warming. While I've spent much of my research career looking at the *global* impacts of climate change, I fully understand that people want to see the *local* impacts. If people are going to understand what is really at stake, scientists have to find new ways to communicate the science, using data, images, and computer scenarios that convey more completely what climate change really looks like—both now and in the future. Beachfront property is only the tip of the iceberg.

This book is written with precisely that goal in mind. It's a book about climate science and climate scientists, but ultimately it lays bare the true stakes of climate change. It illustrates that doing nothing and remaining complacent are tantamount to accepting a future forty years down the road in which your town, your neighborhood, and even your backyard will not look the same. It is not an exaggeration when I say that no place on the planet will look the same forty years down the road if climate change continues. All weather is local, and as you'll see, in the future all climate change will be local, too.

# THE WEATHER OF THE FUTURE

THE WEATHER OF THE FUTURE

PART I

# YOUR WEATHER IS YOUR CLIMATE

# 1

# CLIMATE AND WEATHER TOGETHER

**It's better to build dams than to wait for the flood to come to its senses.**

—Mark Twain

The images were stark: a foreboding gray sky overhead, a turbulent river churning by in the background, and throngs of people—men and women—racing against time to save their town. With the temperature below freezing and snow from a late spring blizzard swirling around them, the volunteers worked hurriedly but efficiently to build makeshift levees, using millions of sandbags. Stuffed into snow boots and down coats, sons and daughters, mothers and fathers, grandmothers and grandfathers tossed bags weighted with sand to each other, each bag moving along to the next person, until at last the bag took its place standing guard alongside the swollen river.

This scene took place on the banks of the Red River during March and April 2009. As the late-season storm swept through, hydrologists at the North Central River Forecast Center warned that the Red River of the North, which runs through the towns of Fargo, North Dakota, and neighboring Moorhead, Minnesota, would crest at 43 feet: 24 feet above flood stage. The situation was tense for days, with the water rising at a seemingly unrelenting rate, but the communities along the river were equally unrelenting. They bagged and tossed around the clock, working in shifts in the frigid air to try to avoid a local catastrophe. People who could not pitch in with the actual bagging helped in other ways, making food, watching kids—banding together to do the work that everyone knew needed to be done. In the end, the job required 3.5 million sandbags and more than 350,000 cubic yards of dirt. Friends and neighbors as well as complete strangers had come together to build a makeshift levee that stretched more than 20 miles.

The Red River flooding of 2009 resulted in a community-wide effort to sandbag, thanks to flood forecasts that provided lifesaving information. Communities along the Red River prepared for more than a week as the U.S. National Weather Service continuously updated the predictions. The entire community around the river as well as state and local authorities came together to use the predictive information as effectively as possible. As the data changed and the severity of the problem rose, people did not sit around hoping that good intentions were enough; they came together to protect their future—even though there was uncertainty as to what exactly the river would do.

The Red River eventually crested at 40.82 feet. During the prolonged flooding, the river was above flood stage for sixty-one days. The U.S. Army Corps of Engineers alone estimates that it spent about $30 million to prevent more than $2 billion in damage. And as bad as the flooding was, the worst fears of the forecasters never came true.

Even so, hydrologists from the National Weather Service said this was the highest crest in more than 120 years of records on the Red River of the North. From the initial point of melt to the peak in the Fargo-Moorhead area, they said it was the fastest a flood like this had ever occurred. The speed, along with the 20 inches of snow that fell, overwhelmed the forecast models; this is why hydrological engineers—studying the flows and peaks of the flooding—are working to refine their forecasting methods. As they work to improve the flood forecast, the Army Corps of Engineers is developing a plan for permanent flood protection for the communities that faced record flooding—because if one thing is certain, there will be a next time.

———

Make no mistake: global warming increases the likelihood of floods such as the Red River flood. This brings me to a central question: if you know a flood is coming, are you going to wait until the water is at your door or are you going to run to the closest riverbank and start pouring sand into a bag?

Global warming has been called the "perfect problem"—perfect in the sense that it's hard to see and challenging to solve. It's hard to see because its signals elude most of our evolutionary panic buttons, save one—our analytical minds. Climate scientists may have built models and issued forecasts, which include mass extinction, submerged coastlines, and chronic food and water shortages; but look outside your window, and there is no sign of a storm fitting that description.

Psychologists say that humans are genetically wired to respond to palpable threats like a stampede of wild elephants or a gun at the back of the head. It's the abstract dangers, the ones we face in the distant future, like global warming, that are tough to wrap our arms around. I get that. I understand that looking at a forecast map for the year 2100, even with the chance of a global average temperature increase of 11°F and a 3-foot rise in global sea level, doesn't set off the requisite alarm bells. And I understand why global warming ranked at the bottom of a list of twenty national priorities in a recent poll by the Pew Research Center.[1] According to the Pew study, our collective list of concerns goes like this: the economy, jobs, terrorism, Social Security, education, energy, Medicare, health care, deficit reduction, health insurance, helping the poor, crime, moral decline, the military, tax cuts, environment, immigration lobbyists, trade policy, and global warming, in that order.

This isn't to say Americans aren't concerned about global warming. Several polls have made it clear that Americans get it; a majority of Americans now feel that global warming is real and that it's caused by human activities. But their concern has done little to alter how we prioritize the risks that global warming poses. Global warming seems less urgent than things staring us in the face. Ultimately, last is still last.

Psychologists chalk up the last-place finish to all the ways that global warming fails to connect with our emotions, our experiences, and our memories. For one, psychologists point to the fact that people have a "finite pool of worry." It's impossible to sustain concern about global warming when other worries, like an economic collapse or a home foreclosure, dive into the pool. Another issue, called the *single-action bias*,[2] is the human habit of taking just one action in response to a problem in situations where multiple solutions are required. For instance, buying your first compact fluorescent lightbulb or using a recycled bag seems to reduce or remove the feeling of worry or concern.

In essence, we aren't fully capable of processing global warming in the traditional human way. So we need to find a new way to look at it, a new way to understand it and break it down.

The traditional human way works something like this. According to cognitive psychologists, we have two different systems for processing risks.[3] One system is analytical. It involves evaluating data and statistics to come up with a careful internal cost-benefit analysis. It's all science. The other system is emotional and drawn from deep personal experience and human memory. This system processes the risk and converts it into a feeling. It makes a situation personal and immediate, and that is why it works quite well in the case of stampeding wild elephants or a gun at the head. These two systems are capable of describing the same event very differently. Research suggests that although the two processing systems operate in parallel, they are both more effective when they're able to interact. And in cases where the outputs from the two processing systems disagree, our emotions and memories usually win. The gun will always trump the numbers.

Or if the gun doesn't trump the numbers, it messes with them. Take, for example, the stock market. It's a classic example of the daily battle between reason and emotion. Data and statistics are fundamental to determining whether or not to buy or sell, but emotions clearly play a role, even when you least expect it. A paper published in the *Journal of Finance* in 2003 found a positive relationship between morning sunshine outside the stock exchange and market index stock returns that day at twenty-six stock exchanges internationally from 1928 to 1997.[4] So much for strictly rational price-setting, and a strong statement about the powerful influence of the weather.

Global warming has a lot of similarities with the stock market. The long-term temperature trend, like the long-term performance of the market, is up. But weather, like day-trading individual stocks, is highly volatile. And like the stock market, global warming is a textbook example of how a disconnect between the analytical and the emotional processing systems often results in a pretty lousy risk assessment. Your brain, after careful analytic consideration, is telling you that of course long-term drought, mass extinctions, and a rising sea level are serious concerns. But your gut just isn't feeling it. It's too far off, too impersonal.

Consequently, many of us are still struggling to see global warming. In fact, when asked to come up with a single, specific image of what global warming looks like, 74 percent of poll respondents see only one thing: melting ice. Although nearly six in ten think global warming is making weather events like droughts and storms more frequent, far fewer connect global warming with specific recent events. In their own personal experience, only 43 percent say weather patterns in the county where they live have become increasingly unstable over the past three years. Experience plays a large role in judging risk. But most of us, especially those in the younger generation, do not yet have experiences that we associate with the threat posed by climate change and cannot bring examples, good or bad, to the table. In fact, our brain is wired to assume that the future will be similar to what we have experienced so far.[5]

Yet having worked at The Weather Channel, I was continually awestruck by the extent to which people rallied around a weather forecast, whether it involved sandbagging in advance of the Red River flood or evacuating in advance of Hurricane Gustav. There's something inspiring about the way communities can pull together under extremely challenging circumstances. We're clearly quite good at processing the risks associated with extreme weather, and this is why it's so important for people to understand that their weather is their climate. Climate and global warming need to be built into our daily weather forecasts because by connecting climate and weather we can begin to work on our long-term memory and relate it to what's outside our window today. If climate is impersonal statistics, weather is personal experience. We need to reconnect them.

———

To understand how we can link climate and weather, it helps to explain why they aren't linked now. The short answer is time.

As climate forecasts and weather forecasts have evolved, they have been separated in the public mind because *weather* is concerned with the immediate whereas *climate* is more focused on the long term. We watch a weather report on Sunday night because we want to know what to expect during the week ahead. The climate forecast, which deals in timescales of months and years, often feels too remote and intangible (unless of course, real estate is involved). We might hear that scientists think this winter will be warmer or this summer will be hotter, but we wait to pass judgment until we can experience it for ourselves. Just as our brain is hardwired to perceive threats that are most immediate to us, we are hardwired to devote more energy to caring about the weather than to caring about the climate.

This separation between weather and climate has been reinforced by the national and local news media, which regularly devote a segment to forecasting tomorrow's weather, but rarely say anything about the climate forecast. It's not that the information isn't available; it's that the way the practice has evolved, we don't expect a climate forecast from our news outlets. As a result, we tend to separate the broad concepts of weather and climate—to see them as vastly different ideas when in reality the only big difference between them is time.

Your daily weather forecast is a function of what is happening in the atmosphere right now. We use the conditions of today (humidity, temperature, wind speed, atmospheric pressure, etc.) to help predict the weather of tomorrow. Meanwhile, climate forecasting gives a broader context to the weather we are currently experiencing. And that context is critical. It is also evolving as a result of greenhouse gas emissions. Think of your daily weather forecast and then average it over time and space—that's roughly what a climate forecast is communicating.

Because weather forecasting and climate forecasting focus on different timescales, their goals are not the same. Whereas weather forecasting is meant to tell you what to expect when you step outside in the morning, climate forecasting is focused on broad trends over time. Will there be a drought next summer? What is the risk of wildfire for the West? Will El Niño appear next year? Will the weather be hotter in 2050? In other words, although I can't tell you whether it will be raining on March 1, 2050, in Fargo, North Dakota, I can say that March, on average, will be warmer and that rainfall, on average, will be more intense.

But despite their different timescales, climate and weather forecasts are focused on achieving a similar result: the means to predict the future. Of course, the question then becomes what do we do with it. The weather forecast is so ingrained in our existence that we know very well how to act on it. If we hear on the radio in the morning that it's going to rain, we carry an umbrella. If we hear that the temperature is going to be unseasonably cool, then we pack a sweater. By definition, weather is a timescale we can't stop. With a weather forecast, we're working strictly on our defense.

However, with the climate forecast the necessary actions are not as straightforward, and this highlights some of the basic philosophical differences between weather and climate. I've come to view a long-range climate projection as an *anti-forecast* in the sense that it forecasts something you want to prevent. Think back to the Red River flood. Until now, we've been able to view extreme weather like flooding as an act of God. But science tells us that, owing to climate change, such floods will happen more often and we need to be prepared for them. I say that a climate forecast is an anti-forecast because it is in our power to prevent the forecast from happening. It represents only one possible future that could happen if we continue to burn fossil fuels as business-as-usual. The future is ultimately in our hands. And the situation is urgent because the longer we wait, the more climate change works its way into the weather, and once it's in the weather, it's there for good.

We are currently in a race against our own ability to intuitively trust what science is telling us, assess the risk of global warming, and predict the future. So when we look at a climate forecast out to 2100 and see temperatures upward of 11°F warmer and sea level 3 feet higher, we need to assess the risk as well as the different solutions necessary to prevent these outcomes. The challenge is to reduce greenhouse gas emissions, replace our energy infrastructure, and adapt to the warming already in the pipeline. And this is the complicated part.

By responding to and trusting the climate forecast, we will prevent it from coming true. Ninety-two percent of those surveyed in a Yale/George Mason poll said the nation should act to reduce global warming. In other words, the overwhelming majority of Americans think we should trust the long-term forecast. But 51 percent of Americans said that although we have the ability to stop global warming, they weren't sure if we actually would. They weren't convinced we'd be able to see and act on the forecast of global warming as the residents of Fargo, North Dakota, saw and acted on the flood forecast. For the people in Fargo, the risk was personal and the forecast was lifesaving.

Most Americans believe that we will not take steps to fix climate change until after it has begun to harm us personally. Unfortunately, by that point it will be too late. The climate system has time lags. And those time lags mean that the climate system doesn't respond immediately to all the extra greenhouse gases in the atmosphere. So, by the time you see it in the weather on a daily basis, it's too late to fix climate change. For most people, the fact that there is uncertainty surrounding the future threat of climate change means we should hold off on any expensive fixes—specifically, actions aimed at reducing greenhouse gas emissions from smokestacks and tailpipes—until we know more. Yet the people of Fargo, North Dakota, didn't wait to see if their town would be flooded; instead, they saw the forecast and started sandbagging. They knew instinctively that if you wait until the water is up to your waist, it's already too late.

# 2

# SEEING CLIMATE CHANGE IN OUR PAST

**It is not the strongest nor most intelligent of the species that survive; it is the one most adaptable to change.**
—Charles Darwin, *On the Origin of Species*

Here's something that most climate scientists won't tell you about climate change: the Earth is going to be fine. As the history of climate change in this chapter shows, the Earth has gone through periods of warming and cooling in the past, and still it remains here. Unfortunately, you can't say the same for the species that occupy Earth—including us. The fact that the Earth will be fine doesn't necessarily mean that the human race will. The last 10,000 to 20,000 years have witnessed a period of dramatic growth in human civilization. Indeed, our growth during this time is unique among all species, but it has been highly dependent on the overall consistency of the climate.

In order to make predictions about the man-made climate change of the future and understand just how high the stakes are, we must first look to the natural climate change of the past. We humans see ourselves as highly adaptable creatures; indeed, whether or not we can endure the coming climate change hinges on our adaptability. However, this is not a given. As a species, we have never been forced to adapt to a global increase in temperature like the one we currently face. If climate change does indeed occur, the future for humans will become a lot less certain—we will be much like other animals before us who proved unable to adapt to changing climates.

Take the woolly mammoth, the unofficial mascot of the ice ages. To see a woolly mammoth is to see a climate that no longer exists, and it's been this way ever since people first started finding mammoth fossils.

Weighing 20,000 pounds, and with tusks 16 feet long, mammoths must have looked quite striking 20,000 years ago, as they strolled around what is now downtown Los Angeles. Obviously, this region was very different 20,000 years ago, and the climate of Earth was very different as well. It's a time scientists refer to as the Last Glacial Maximum (LGM). Major areas of the Earth were locked in relentless winter, covered in massive sheets of ice that grew in frigid strongholds to the north. Los Angeles was not covered by ice, but it was definitely influenced by the cold elsewhere. Forests, fields, and even mountains were no match for these vast sheets of ice, and they had the battle scars to prove it. The ice was voracious, drinking up the oceans and drawing down the sea level by almost 400 feet. But all this was not a problem for the woolly mammoth—quite the contrary. Woolly mammoths were well adapted to the cold climate of the LGM, with shaggy hair more than 3 feet long to protect them from frigid winds and a 3-inch layer of blubber to keep them warm.

This vast stretch of ice, known as the Laurentide ice sheet, buried what is today Canada, New England, the Midwest, and parts of Washington, Idaho, and Montana under a layer of ice more than 1 mile thick.[1] Just south of this vast ice sheet stretched a treeless tundra that was equally expansive. This was the summer home of the woolly mammoth. The mammoths nibbled on the coarse tundra grasses with their perfectly adapted, but probably not pearly white, teeth. Woolly mammoth teeth, in fact, were about 6 inches square, the biggest grinding teeth in the animal kingdom. Those teeth were perfectly suited to their mammoths' ice age vegetarian diet.

Evidence of mammoths has been found throughout the northern hemisphere. Early humans settling along the North Sea coast, sometime between 6,000 and 8,000 years ago, also encountered skeletal remains of the woolly mammoth. The low sea level at the end of the last ice age provided an exposed shelf between the Netherlands and England that the mammoths roamed freely across. When the sea level rose again, rough surf would have crushed the exposed skeletons of the woolly mammoths, but their teeth were tough as nails, and these would have survived and eventually washed up along the shore. All along the North Sea, storm waves would have tossed woolly mammoth fossils up onto the beach, like seashells, for Vikings to find.

I can't imagine what it must have been like finding a woolly mammoth tooth along the shore all those years ago. Such a discovery most certainly raised some tricky questions that would have been tough to answer at the time. An animal with teeth 6 inches wide? There was nothing walking around the North Sea coast at that time with teeth 6 inches wide. Had the North Sea settlers been able to ask their Stone Age ancestors who lived through the ice age 10,000 years earlier, these ancestors would have been able to explain everything. They had, in fact, hunted the woolly mammoth. But lacking a time machine, the early Viking settlers had to come up with their own story to explain the existence of these very large teeth.

On the basis of the size of the teeth, the Vikings calculated that the animal must have been more than 70 feet tall. In tribute they named their new home, in what is today Denmark and Germany, "Land of the Giants." They assumed the woolly mammoths were the children of an enormous ice giant to the north who had once ruled all of Scandinavia. The legend went on to say that when the ice giant was killed, his blood made the sea level rise and drowned all his furry children with the big teeth. This explanation, preserved today in Icelandic sagas, is the earliest recorded notion pointing to the existence of an ice age, and actually it's not that bad an explanation for what happened.

————

The early Vikings were some of the first people in recorded history to try to understand how and why climate change occurred. Perhaps one of the greatest misconceptions about climate change is the notion that studying it is something that began in the twentieth century. In fact, many of the first important discoveries about global warming were made during the 1800s.

A lot of people are surprised to learn that scientists have been working on the problem of global warming for well over 100 years. The key difference in the beginning, though, was that the scientists weren't studying humanity's role in the process: they were trying to understand something that for religious and cultural reasons was a dangerous idea at the time: perhaps the climate on Earth had not always been the same.

The base of climate science today comes from work that was done by these visionaries of the nineteenth century. What's so impressive about these pioneers is that they were able to see climate in ways no one had ever seen it before. They were trying to find answers to fundamental questions: Why is the sky blue? How old is our planet? Why were woolly mammoth bones popping up in the La Brea Tar Pits in Los Angeles? And so on. These scientists were starting from scratch in building a body of evidence about the Earth's climate. They had to frame the questions, devise the equipment, and then perform the experiments to come up with reproducible answers. Getting the planet to share its past is like pulling teeth. But, as it turns out, teeth had a lot to say.

In the 1800s when scientists once again began finding 6-inch teeth scattered across North America and Europe and into Siberia, they wanted to do something a little better than a Viking myth about an ice giant. They wanted to use the tools of science to build a rigorous explanation that could stand the test of time. Ironically, they ended up proving that the Vikings weren't too far off, at least with regard to the giant ice age.

In 1837 Louis Agassiz, a Swiss scientist, stood up before his colleagues at a conference in the Swiss town of Neuchâtel to present a theory suggesting that the Earth had indeed experienced an ice age. Like many others in his day, he had observed the glaciers of his native Switzerland and noticed the marks that these glaciers left behind: rocks with scratches and scars, mounds of debris called moraines that had been pushed up by glaciers, deep valleys, signs that large boulders had been carried long distances. Agassiz came to realize he was seeing classic signs of a process known as glaciation in places where there were no glaciers to be seen.

Agassiz was going up against an explanation that had come from the Bible. At the time, it was widely believed that *The Great Flood* was the only event with the power to do such heavy lifting.[2] The story of Noah's flood, with just a slight tweak, received almost unanimous support from the scientific community. The tweak had been provided by the great English geologist Charles Lyell. And it was required in order to overcome an inconsistency in the story. Lyell's revision to The Great Flood was his suggestion that the big boulders dropped off in strange places had, in fact, been transported by icebergs.

But that still left the issue of the strange scars on the rocks. Interestingly, plenty of local villagers at the time had already come to their own conclusions about these scars. Having grown up among glaciers, they didn't need a scientist to explain the origins of the strange scars. Throughout the towns and villages of Switzerland, it seems many people had already been convinced that the scratches and scars were the result of a flood of ice, not a flood of water as Lyell had suggested. They, in fact, had already come to accept the theory of a great ice age, just like the Vikings before them.

Despite the lukewarm reception of his presentation at the Swiss Society of Natural Sciences in Neuchâtel in 1837, Agassiz persisted.

In 1840, he even published a book called *Studies on Glaciers*.[3] With the help of numerous colleagues who had been convinced by his evidence and by the clarity of his argument, Agassiz fought hard to convince skeptics who clung to the theory of The Great Flood. Eventually, the overwhelming strength of the evidence won out. In the end, Agassiz had proved that there was a period of time when large areas of the Earth had been covered by ice sheets. Like the Swiss villagers, he had come up with the simplest and most consistent explanation. The ice age, he said, reached its maximum about 20,000 years ago, and then gave way to an eventual warming.

Let's go back to the woolly mammoth for a moment. The rise and fall of the woolly mammoth is linked to the rise and fall of the ice ages. The rise began around 300,000 years ago, as the Earth underwent a transition to a cooler climate. The peak of the last glacial period, the LGM, was about 20,000 years ago. After that, over a span of about 12,000 years, much of the ice melted, the sea level rose almost 400 feet, and the temperature rose about 11°F. Fossil evidence suggests that at the peak of the LGM, woolly mammoths could be found across Europe, Asia, and North America. They were so well adapted to the cold that during the last ice age, parts of Siberia may have had an average population density of about sixty woolly mammoths to every 40 square miles. But then, as the climate changed around them, they simply died out. As scientists processed the significance of this connection, they had to invent a word to explain the phenomenon. The word is *extinction*.

The woolly mammoth, that icon of the ice age, also became an icon of extinction. Before their extinction was recognized, no one had supposed that a robust species could simply disappear. So Louis Agassiz will always be credited not only with his theory of a "great ice age" but also with discovering extinction.

Over the years, Agassiz's theory of the ice age needed to be refined. The ice sheets were not as large as he had thought, and the ice age didn't arrive as suddenly as he had thought. Most important, there wasn't just one great ice age. In Scotland, plant fragments were found sandwiched between layers of glacial deposits. It became increasingly obvious that there had been not just one ice age but several large glaciations, one following another, separated by warm periods. Scientists came to understand that the Earth actually moved into and out of ice ages. And with that amazing discovery, a new crop of scientists began to work on a new theory they called *climate change.*

————

To grow a continental-scale ice sheet you need low temperatures. That much was clear. What wasn't so clear was how the temperatures had been lowered enough to permit the growth of ice on such a massive scale.

One important hypothesis of how the planet regulated its temperature was put forth by the French mathematician and physicist Joseph Fourier in 1824.[4] As a physicist, Fourier was interested in understanding some basic principles about the flow of heat around the planet. Specifically, he wanted to use the principles of physics to understand what sets the average surface temperature of Earth. It made perfect sense that the sun's rays warmed the surface of the Earth, but this left a nagging question: when light from the sun reaches the surface of the Earth and heats it up, why doesn't the Earth keep warming up until it's as hot as the sun? Why is the Earth's temperature set at roughly 59°F—the average temperature at its surface?

Fourier reasoned that there must be some balance between what the sun sends in and what the Earth sends back out, so he coined the term *planetary energy balance,* which is simply a way of saying that there is a balance between energy coming in from the sun and energy going back out to space. If the Earth continuously receives heat from the sun yet always has an average temperature hovering around 59°F, then it must be sending an equal amount of heat back to space. Fourier suggested that the Earth's surface must emit invisible infrared radiation that carries the extra heat back into space. Infrared radiation (IR), like sunlight, is a form of light. But it's a wavelength that our eyes can't see.

This was a good idea, but when he actually tried to calculate the planet's temperature using this effect, he got a temperature well below freezing. So he knew he must be missing something. To arrive at 59°F, the Earth's average temperature, Fourier realized that he needed the atmosphere to pick up the slack. And he discovered a phenomenon he called the *greenhouse effect*, a process whereby the gases in the Earth's atmosphere trap certain wavelengths of sunlight, not allowing them to escape back out to space. Like the glass in a greenhouse, these *greenhouse gases* let sunlight through on its way in from space, but intercept infrared light on its way back out.

In 1849, an Irish scientist, John Tyndall, was able to build on this idea. He had become obsessed with the glaciers he climbed while visiting the Alps on a vacation. Like many other scientists at the time, he wanted to understand how these massive sheets of ice formed and grew. He applied his personal observations of glaciers in the laboratory in 1859, when, at the age of thirty-nine, he began a series of innovative experiments.

Tyndall was intrigued by the concept of a *thermostat*. We know thermostats today as devices that regulate the temperature of a room by heating or cooling it. Tyndall devised an experiment to test whether the Earth's atmosphere might act like a thermostat, helping to control the planet's temperature. He reasoned that it might help explain how ice ages had blanketed parts of the Earth in the past.

For his experiment, Tyndall built a device, called a spectrophotometer, which he used to measure the amount of radiated heat (like the heat radiated from a stove) that gases such as water vapor, carbon dioxide, or ozone could absorb. His experiment showed that different gases in the atmosphere had different abilities to absorb and transmit heat. Some of the gases in the atmosphere—oxygen, nitrogen, and hydrogen—were essentially transparent to both sunlight and IR, but other gases were in fact opaque: they actually

absorbed the IR, as if they were bricks in an oven. Those gases include carbon dioxide ($CO_2$) and also methane, nitrous oxide, and
water vapor. These greenhouse gases are very good at absorbing
infrared light. They spread heat back to the land and the oceans.
They let sunlight through on its way in from space, but intercept
IR on its way back out. Tyndall knew he was on to something. The
fact that certain gases in the atmosphere could absorb IR implied
a very clever natural thermostat, just as he had suspected. His top
four candidates for a thermostat were water vapor, without which he
said the Earth's surface would be "held fast in the iron grip of frost";
methane; ozone; and, of course, carbon dioxide.[5]

Tyndall's experiments proved that Fourier's greenhouse effect
was real. They proved that nitrogen (78 percent) and oxygen (21
percent), the two main gases in the atmosphere, are not greenhouse
gases, because a molecule of each of these elements has only two
atoms, so it cannot absorb or radiate energy at IR wavelengths.
However, water vapor, methane, and carbon dioxide, each of which
is a molecule with three or more atoms, are excellent at trapping
IR radiation. They absorb about 95 percent of the long-wave or IR
radiation emitted from the surface. So, even though there are only
trace amounts of these gases in the atmosphere, a little goes a long
way toward making it really tough for all the heat to escape back
into space. In other words, greenhouse gases in the atmosphere
act as a secondary source of heat, in addition to the sun. And the
greenhouse gases provide the additional warming that Fourier
needed to explain that average temperature of 59°F.

Thanks to Tyndall, it is now accepted that visible light from the
sun passes through the Earth's atmosphere without being blocked
by $CO_2$. Only about 50 percent of incoming solar energy reaches
the Earth's surface: about 30 percent is reflected by clouds and the
Earth's surface (especially in icy regions), and about 15 percent is
absorbed by water vapor. The sunlight that makes it to the Earth's
surface is absorbed and reemitted at a longer wavelength, IR, that
we cannot see, like heat from an oven. Carbon dioxide (like other

heat-trapping gases, such as methane and water vapor) absorbs the IR and warms the air, which in turn warms the land and water below it. More carbon dioxide means more warming. This is where the concept of a natural thermostat becomes very powerful—mess with the amount of $CO_2$ in the atmosphere, and you're resetting the thermostat of the planet.

The idea was good, even profound, but the term *greenhouse effect* was not entirely accurate. Real greenhouses stay warm without a heater because the sun's rays shine in, warming the inside of the greenhouse, and the glass keeps the heat from escaping. But in reality the atmosphere is much more sophisticated than a greenhouse. Fourier had figured out something very important. He had figured out that the sun is not our only source of heat. The atmosphere, in fact, is a very powerful backup generator. This was yet another discovery on the road to understanding the relationship between temperature and carbon dioxide, a relationship that turns out to have profound implications for our climate.

———

Svante Arrhenius (1859–1927), a Swedish physicist and chemist, was another scientist who was smitten with ice ages. He took Tyndall's thermostat mechanism and ran with it, exploring whether the amount of $CO_2$ in the atmosphere could be fiddled with by an event such as a volcanic eruption. According to Tyndall's experiments, the additional carbon dioxide released by the volcano could conceivably raise the Earth's temperature, and Arrhenius wanted to see if that was actually true.

We refer to events or processes that result in changes to the climate as *forcings*. A volcanic eruption is an example of a natural forcing. A forcing can often result in an amplification (positive) or a reduction (negative) in the amount of change and often comes hand in hand with a *feedback*—a situation where some effect causes more of itself. In other words, if a forcing is the event that creates change, then the feedback amplifies that change. But keep in mind that a positive feedback is not positive in the sense of being good. *Positive* refers specifically to the direction of change, not to the desirability of the outcome. A negative feedback tends to reduce or stabilize a process, whereas a positive feedback tends to increase or magnify it.

Maybe, Arrhenius thought, this positive feedback mechanism was responsible for plunging the planet into an ice age. If the atmosphere were to dry out for some reason, the decreasing water vapor would hold less heat and the Earth would cool. Since cooler air holds less water vapor, the atmosphere would tend to dry more, amplifying the cooling. In addition, cooler temperatures would generally lead to increases in snow and ice, and so to yet another positive feedback. When snow and ice cover a region, such as the Arctic or Antarctica, their white, light-reflecting surface tends to bounce sunlight back out to space, helping to further reduce temperature. If regions covered by snow and ice expanded over more of North America and Europe, the climate would cool further while also increasing the ice sheets. Start with a drop in carbon dioxide, continue with a drop in temperature, add some snow and ice, and you've made an ice age.

Arrhenius thought his theory was quite solid, but he wanted to prove it mathematically. So he set about a series of grueling calculations that attempted to estimate the temperature response of changing levels of carbon dioxide in the atmosphere. These may have begun as "back of the envelope" calculations, but in 1896 he was confident enough to publish the work for his colleagues to read.[6] The end result of all of it was one simple number: 8°F.

That number represented roughly how much Arrhenius thought the Earth's average temperature would drop if the amount of $CO_2$ in the atmosphere fell by half. Once you factor in the positive feedbacks of water vapor, snow, and ice, an ice age seemed like a reasonable outcome. The only thing Arrhenius still needed was a mechanism for tinkering with atmospheric carbon dioxide, turning down the natural thermostat. And that is what led, in part, to the discovery of the carbon cycle.

Arrhenius asked a colleague, Arvid Högbom, to help him figure out how much carbon dioxide levels in the atmosphere might be able to change. Högbom had compiled estimates of how carbon dioxide flows through various parts of the planet, including emissions from volcanoes, absorption by the oceans, and so forth. This carbon cycle is a fundamental concept that is hugely important. If carbon dioxide really was the natural thermostat that scientists had been searching for, then the next crucial step would be to figure out how $CO_2$ cycles into and out of the ocean, the land, the atmosphere, and living matter such as plants and trees.

It turns out that carbon (the C in carbon dioxide) has the ability to cycle among a few different reservoirs. Relatively small amounts of carbon reside in the atmosphere, the ocean surface, and vegetation. A slightly larger amount is held in soils, and a much larger amount resides in the deep ocean. The biggest reservoir can be found in rocks and sediments. Carbon takes different chemical forms in different reservoirs. In the atmosphere, it is the gas carbon dioxide ($CO_2$).

The carbon cycle can be thought of, metaphorically, as a kind of reincarnation. This cycle is the great natural recycler of carbon atoms. The same carbon atoms in your body today have been used in countless other molecules for millions, even billions, of years. The wood burned in a fireplace last winter produced $CO_2$ that found its way into a tomato plant this spring. The borders are wide open and carbon cycles easily cross different zones. The atoms pair up, get into various substances for a while, come out of those, and go somewhere else—it is a continuous and ongoing cycle.

Here's a carbon cycle scenario. In phase one, volcanoes and hot springs transfer carbon from deep below the Earth's crust to the atmosphere. In phase two, the carbon dioxide is scrubbed from the atmosphere by a process called *chemical weathering*. Basically, when it rains, the rainwater combines with $CO_2$ in the atmosphere to form a weak acid, carbonic acid. That weak acid falls as rain and then chemically reacts with rocks, releasing carbon, which eventually makes its way into the ocean, where it is locked up in the shells of marine plankton.[7] After dying, the marine plankton eventually sink to the bottom and turn into rocks.

Here, the scenario gets really interesting. Experiments show that rates of chemical weathering are influenced by three environmental quantities: temperature, precipitation (rain and snow), and plant matter. Temperature, precipitation, and vegetation all act in a mutually reinforcing way to affect the rate of chemical weathering. The higher the temperature, the faster a rock is broken down by chemical weathering. Higher precipitation raises the level of groundwater held in soils and combines with $CO_2$ to form carbonic acid and more rapidly drive the weathering process. Remember that temperature and precipitation are linked; the amount of water vapor that air can hold rises with temperature. Likewise, the amount of vegetation is closely tied to temperature and precipitation. More rainfall means more vegetation, and more vegetation means more carbon stored in the soil.

So, carbon becomes the secret ingredient in adjusting the natural thermostat and changing the Earth's climate. The beauty of this mechanism is that it's a big loop. On the one hand, the speed of chemical weathering is tuned to the state of the Earth's climate. On the other hand, the climate is tuned to the rate at which $CO_2$ is pulled out of the atmosphere by chemical weathering. This is an example of a very sophisticated feedback loop.

Ultimately, chemical weathering is the most likely explanation for Earth's habitability over most of the 4.6 billion years of its existence. Any factor that heated Earth during any part of its history caused chemical weathering rates to increase. This increase, in turn, drew $CO_2$ out of the atmosphere at faster rates, and eventually resulted in a cooling to offset the warming. On the flip side, any factor that cooled Earth set off the opposite sequence of events. Chemical weathering constantly acts to moderate long-term climate changes by adjusting the $CO_2$ thermostat as needed. If positive feedbacks help push our climate into an ice age, chemical weathering helps to push us out of one.

As a result of chemical weathering, most of Earth's carbon is tied up below the surface in rocks and pools, including coal, oil, and natural gas. But now, of course, we humans are taking the coal, oil, and natural gas out of the ground and burning it, transferring long-stored carbon to the atmosphere. Nature's history tells us what to expect.

––––––––

We tend to think of man-made global warming as a modern concept, something that has come into vogue in the last twenty years or so, but in reality this idea is more than 100 years old. As noted above, the notion that the global climate could be affected by human activities was first put forth by Svante Arrhenius in 1896. He based his proposal on his prediction that emissions of carbon dioxide from the burning of fossil fuels (i.e., coal, petroleum, and natural gas) and other combustion processes would alter atmospheric composition in ways that would lead to global warming. Arrhenius calculated how much the temperature of the Earth would drop if the amount of $CO_2$ in the atmosphere was halved; he also calculated the temperature increase to be expected from a doubling of $CO_2$ in the atmosphere—a rise of about 8°F.

More than a century later, the estimates from state-of-the-art climate models doing the same calculations to determine the increase in temperature due to a doubling of the $CO_2$ concentration show that the calculation by Arrhenius was in the right ballpark. The Fourth Assessment Report of the Intergovernmental Panel on Climate Change (IPCC) synthesized the results from eighteen climate models used by groups around the world to estimate climate sensitivity and its uncertainty. They estimated that a doubling of $CO_2$ would lead to an increase in global average temperature of about 5.4°F, with an uncertainty spanning the range from about 3.6°F to

8.1°F. It's amazing that Arrhenius, doing his calculations by hand and with very few data, came so close to the much more detailed calculations that can be done today.

Arrhenius's calculations, however, did have some shortcomings. For example, in estimating how long it would take for the $CO_2$ concentration in the atmosphere to double, he assumed that it would rise at a constant rate. With about 1.6 billion people on the planet in 1895 and with relatively small use of fossil fuels, Arrhenius predicted that it would take about 3,000 years for the atmospheric $CO_2$ concentration to double. Unfortunately, when scientists today factor in the quadrupling of world population since then and the increasing demand for energy, doubling is now projected before the end of this century unless substantial cutbacks in emissions are adopted by nations around the world. So, technically, Arrhenius was off by about 2,800 years. (Another of his doubtful predictions was that he firmly believed a warmer world would be a good thing.)

In Arrhenius's time, the impacts of global warming were mainly left to future investigation—the majority of scientists still needed to be convinced that the concentration of $CO_2$ in the atmosphere could vary, even over very long timescales, and that this variation could affect the climate. Scientists at the time were focused more on trying to understand the gradual shifts that took place over periods a thousand times longer than Arrhenius's estimate: those that accounted for alternating ice ages and warm periods and, in distant times (more than 65 million years ago), for the presence of dinosaurs. They couldn't even begin to wrap their minds around climate change on a human timescale of decades or centuries. Nobody thought there was any reason to worry about Arrhenius's hypothetical future warming, which he suggested would be caused by humans and their burning of fossil fuel. It was an idea that most experts at the time dismissed. Most scientists of the era believed that humanity was simply too small and too insignificant to influence the climate.

Fast-forward to the mid-1950s, and enter Charles David Keeling, a brilliant and passionate scientist who was then beginning his research career at Caltech. Keeling had become obsessed with carbon dioxide and wanted to understand what processes affected fluctuations in the amount of $CO_2$ in the atmosphere. Answering this question required an instrument that didn't exist, the equivalent of an ultra-accurate "atmospheric Breathalyzer." So Keeling built his own instrument and then spent months tinkering with it until it was as close to perfect as he could get at measuring the concentration of $CO_2$ in canisters with a range of values of known concentration.

Keeling tried his instrument out by measuring $CO_2$ concentrations in various locations around California and then comparing these samples in the lab against calibration gases. He began to notice that the samples he took in very pristine locations (i.e., spots where air came in off the Pacific Ocean) all yielded the same number. He suspected that he had identified the baseline concentration of $CO_2$ in the atmosphere; a clear signal that wasn't being contaminated by emissions from factories, farms, or uptake by forests and crops.

With this instrument, called a *gas chromatograph*, Keeling headed to the Scripps Institution of Oceanography to begin what is perhaps the single most important scientific contribution to the discovery of global warming. Keeling was on a mission to find out, once and for all, if $CO_2$ levels in the atmosphere were increasing. He would spend the next fifty years carefully tracking $CO_2$ and building, data point by data point, the finest instrumental record of the $CO_2$ concentration in the atmosphere, generating a time history that is now known by scientists as the Keeling curve.

The Keeling curve is a monthly record of atmospheric carbon dioxide levels that begins in 1958 and continues to today. The instrument Keeling built, the gas chromatograph, works by passing infrared (IR) light through a sample of air and measuring the amount of IR absorbed by the air. Because carbon dioxide is a greenhouse gas, Keeling knew that the more IR absorbed by the air, the higher the concentration of $CO_2$ in the air. Because $CO_2$ is found in very small concentrations, the gas chromatograph measures in terms of parts per million (ppm).

Keeling knew from his travels around California that he needed to make his measurements at a remote location that wouldn't be contaminated by local pollution. That's why he settled on Hawaii. Hawaii's big island is the site of the volcano Mauna Loa, and Keeling set up his $CO_2$ instrument near the top of Mauna Loa. Isolated in the middle of the Pacific Ocean and at more than 11,000 feet above sea level, the top of the Mauna Loa volcano is an ideal location to make measurements of atmospheric carbon dioxide that reflect global trends, but *not* local influences such as factories or forests that may boost or lower the carbon dioxide level within their vicinity. The sensors were positioned so that they sampled the incoming ocean breeze well above the thermal inversion layer; thus the air was not affected by nearby human activities, vegetation, or other factors on the island. Obviously, volcanoes are potentially a big source of $CO_2$, but Keeling took this into account when positioning his instrument, locating it upwind of Mauna Loa's vent and installing sensors to give alerts if the winds shift.

What he found was both disturbing and fascinating, creepy and profound. Keeling, using his Mauna Loa measurements, could see that with each passing year $CO_2$ levels were steadily moving upward. As the years passed and the Mauna Loa data accumulated, Keeling's $CO_2$ record became increasingly impressive, showing levels of carbon dioxide that were noticeably higher year after year. The first instrumental measurements indicated a $CO_2$ concentration of 315 ppm in 1958. The slow rise in its concentration over the first several years was enough to prompt a report from a panel of the President's Science Advisory Council to President Johnson in 1965, indicating that the early prediction that an increase in $CO_2$ could occur was correct and that global warming would indeed be expected to occur. This was the first instance when a document discussing global warming ended up in front of the president of the United States. It would not be the last.

In 2008, just over fifty years after Keeling started his observations, the concentration at Mauna Loa had reached 385 ppm. Keeling's measurements thus provided solid evidence that the atmospheric $CO_2$ concentration was increasing. If anything proved that Arrhenius had been on to something, it was these data.

One of the most striking aspects of the Keeling curve is a small $CO_2$ wiggle that takes place every year. For every little jump up, there is a little dip back down, so that the whole curve looks sawtoothed. This wiggle happens like clockwork and is timed with the seasons. In the northern hemisphere during fall and winter, plants and leaves die off and decay, releasing $CO_2$ back into the atmosphere and causing a small spike. And then during the spring and summer, when plants are taking $CO_2$ out of the atmosphere in order to grow, carbon dioxide levels drop. Hawaii, along with most of the planet's landmass, is situated in the northern hemisphere, so the seasonal trend in the Keeling curve is tracking the seasons in the northern hemisphere. The Keeling curve proved many important things at once. It proved that $CO_2$ levels in the atmosphere can indeed change and that they can change on very short timescales.

Keeling's record was the icing on the cake, and he rightly stands with Agassiz, Tyndall, and Arrhenius among the giants of climate science. He helped prove the reality of global warming by providing the data upon which the pioneering theories of Tyndall and Arrhenius could finally rest. As is the case in research science, Keeling's painstaking measurements have been verified and supplemented by many others. Measurements at about 100 other sites have confirmed the long-term trend shown by the Keeling curve, although no sites have a record as long as Mauna Loa. Other scientists have also extended the Keeling curve farther back in time, using measurements of $CO_2$ in air trapped in bubbles in polar ice and in mountain glaciers. Ice cores collected from Antarctica and Greenland can be used to reconstruct climate hundreds of thousands of years ago, showing that the preindustrial amount of $CO_2$—the level from A.D. 1000 to 1750—in the atmosphere was about 280 ppm, about 105 ppm below today's value. The record indicates that the concentation of $CO_2$ has increased about 36 percent in the last 150 years, with about half of that increase happening in the last three decades. In fact, the $CO_2$ concentration is now higher than any seen in at least the past 800,000 years—and probably many millions of years before the earliest ice core measurement.

Over the past century, the evidence has piled up in support of Arrhenius's explanation of global warming. As the evidence accumulates with each passing year, what was once a fringe hypothesis that sprang from the mind of a single scientist in Sweden is now part of the bedrock of scientific accomplishments. Unfortunately, scientific discoveries are not always good news. And there is a nagging fear among scientists that we'll prove ourselves to be not so different from the woolly mammoth, the symbol of a climate that no longer exists.

# 3

# THE SCIENCE
# OF PREDICTION

**If I have seen further, it is by standing on the shoulders of giants.**

—Sir Isaac Newton

Prediction is an odd thing. Depending on your personality, predictions are a source of either comfort or anxiety. In one broad stroke, they have the power to reassure or destabilize. Predictions often give us an illusion of control in situations that are inherently out of our control. Nothing exemplifies this better than our relationship to weather forecasts. We can't stop the weather, but we can at least prepare for it. Ultimately, this preparation is what the science of prediction—be it climate or weather prediction—is all about.

A certain pleasure comes from knowing that meteorologists are generally right about the forecast, and a certain disappointment comes from finding out they got it wrong. Of course, we are not happy when predictions fail. Over the last fifty years, we have grown accustomed to the idea that the weather can be "predicted," so it feels like violation when a forecast turns out to be incorrect.

A big part of believing predictions like those in this book has to do with trusting and understanding the underlying data and models. Model simulations are the closest thing that scientists have to a crystal ball, and as a result data are the lifeblood of every prediction that weather and climate scientists make. At this point, weather prediction is so ingrained in our lives that we've stopped being skeptical about it. Even though the sun is shining, our experience tells us that we should trust the man or woman in front of the map who's gesturing at swirling shades of green behind it. Unfortunately, the same cannot be said for climate prediction; but as we will see here, the two are really not all that different.

Although weather prediction is now embedded in our psyche, the practice as we know it has been studied for only about 100 years. What we need to understand is that the mechanisms of weather predictions are very similar to those of climate predictions. So if we're comfortable trusting local forecasters' predictions about weather, we should probably think about trusting the predictions coming out of the country's climate laboratories.

———

The modern-day weather forecast originated on the battlefields of World War I. During that war, a young Quaker ambulance driver, Lewis Fry Richardson, fascinated by the possibility of seeing the weather before it happened, laid the groundwork for the daily weather forecasts that we all live by today.[1] Richardson, a true giant in weather forecasting, was also a pioneer in a branch of mathematics called *numerical analysis*. Numerical analysis looks for ways to find approximate solutions to problems that are too complicated to solve. It also serves as a bridge between people and computers. One of the key differences between people and computers is that computers can do arithmetic lightning fast. Humans, on the other hand, can come up with elegant mathematical equations to represent how the world works. Despite their elegance, those mathematical equations are hard to solve, and that's where numerical analysis comes in handy. Without it, computer models would not have been possible.

But before there were computers, there was Richardson. He was committed to the idea of generating the very first weather forecast using seven elegant mathematical equations developed by another giant in the field of meteorology, the Norwegian scientist Vilhelm Bjerknes. By the time of World War I, Bjerknes had come up with equations capable of describing the behavior of the atmosphere. The state of the atmosphere at any point could be described by seven values: (1) pressure, (2) temperature, (3) density, (4) water content, and wind—(5) east, (6) north, and (7) up. In essence, Bjerknes presented Richardson with seven complex calculus problems in need of transformation.

Richardson knew that the differential equations could be approximated and simplified using numerical analysis. And once the equations were simplified, he figured that he should be able to generate a weather forecast for central Europe. To do this, he divided the entire atmosphere into discrete columns measuring about 3° east–west and about 125 miles north–south; this division works out to about 12,000 columns on the surface and five rows in the atmosphere. If he calculated the value of each of the seven variables for each cell in the two columns over central Europe, he figured he'd have the first battlefield weather forecast.

Of course, at that time, Richardson did all his work by hand, in "offices" that can most charitably be described as airy—temporary rest camps with a view of the front lines of the fighting. Computers capable of doing the math were still a far-off dream, so with just pencil and paper, this driver with the Friends Ambulance Unit in France tackled the problem of weather prediction. Richardson himself was the computer. His forecast for central Europe was no small undertaking; he later wrote, "The scheme is complicated because the atmosphere is complicated." Even the simplified procedure required a maddening amount of arithmetic. There was so much arithmetic to be done that calculating a weather forecast just six hours out in time required about six months of work—rather late to be considered an actual forecast.

But Richardson was undaunted, expressing his dream that "someday in the dim future it will be possible to advance the computations faster than the weather advances." Of course, when he wrote these words, Richardson was imagining people doing the calculating. In the not too distant future, artificial computers would easily outrun time and see the future before it happened.

Unfortunately, the time it took to grind out the forecast calculations wasn't Richardson's only problem. The *initial conditions* he used to start the calculation were both incomplete and imprecise. He just didn't have all the observational data he needed to fully represent the physical state of the atmosphere. As a result, the first official weather forecast went down in history as a total bust, and with that bust came one of the cardinal rules of weather and climate prediction: your forecast is only as accurate as your data.

Still, though the forecast itself was off, much of what Richardson proposed was right. And luckily for all of us who count on reliable weather forecasts today, Richardson was brave enough to publish his ideas. However, he had to find his manuscript first—he had lost the sole copy during the Battle of Champagne in April 1917. He discovered it months later under a heap of coal. The book, eventually published in 1922, was called *Weather Prediction by Numerical Process*. And what at first appeared to be nothing more than a failed weather forecast is now widely considered one of the most profound books about meteorology ever written. Richardson had come up with a way to see into the future. But he couldn't do it alone. He needed computers.

Upon returning from the war, Richardson eventually quit meteorology when he realized that his work was being used for military purposes. A committed pacifist, this gentle giant actually destroyed some of his research to prevent it from being used by the military. He spent much of the rest of his life applying mathematics to the understanding of the causes of war. But as time went on, and the field of theoretical meteorology came of age, Richardson's early vision of a weather forecast was fully realized. Computers were being developed, and by the late 1940s the first successful numerical weather prediction was performed at the Institute for Advanced

Study in Princeton, New Jersey. By the 1950s routine weather forecasts were being produced; these used very simple models that did not take into account variables such as radiation and so led to some fairly large errors.

Yet in spite of these shortcomings, the computers proved very effective at predicting the weather, especially as more advanced forms of data collecting fed more accurate information into the models. Today, the North American Mesoscale (NAM) model developed by the National Weather Service—this is the model that The Weather Channel uses for its forecasts—takes about ninety minutes to ingest all the data (those very important initial conditions), and the actual computer calculations that provide the weather forecast out to eighty-four hours (3.5 days) take less than ninety minutes. So, give a model three hours, and it'll give you the weather for the entire country for the next three days.

———

As weather forecasts became more routine and forecasters' skill increased, scientists began to look for a new challenge; they began to look farther out in time. The goal was to build a model that represented the climate system. This was no small task. Weather models are concerned only about what's happening in the atmosphere. The atmosphere has a memory of roughly one week. That's why your local weather forecast goes out only about a week.

Climate models, however, needed to include much more. Scientists had to connect their mathematical version of the atmosphere to mathematical versions of the oceans, the land surface, and sea ice and biology. This was a vast expansion of weather prediction. And so, in the late 1940s, scientists, many of them meteorologists, set out to derive the mathematical equations that would describe the rest of the planet. They were building a computer model that would serve as a planetary stunt double. It would be an entirely new way of looking even farther into the future.

Under the direction of Joseph Smagorinsky at the U.S. Weather

Bureau in Washington, D.C., the work started with basic physics equations of fluids and energy, and then kept building from there. Syukuro Manabe, a Japanese meteorologist, arrived in the United States from Tokyo University in 1958 to help Smagorinsky. He began work on an atmospheric model that would include the basics: winds, rain, snow, and sun. He and Smagorinsky also included the greenhouse effect caused by both carbon dioxide and water vapor. This would allow them to eventually test what the increased carbon dioxide would do to the climate system.

In the meantime, building this "twin Earth" required understanding the nitty-gritty of how the world works. Manabe found himself in the library researching topics such as how different soils absorb water. By 1965 he and Smagorinsky had developed a three-dimensional model, which solved the basic equations for the atmosphere and was simple enough that the equations could be calculated efficiently. Still, it's important to keep in mind that this early model, and others, had no geography: no land and no oceans. Everything was averaged over bands of latitude, with continents and oceans mixed together to form a swamp that exchanged moisture with the atmosphere above it but was unable to absorb heat. All in all, the atmosphere generated by these models looked decent. The model output showed a realistic layered atmosphere, as well as a zone of rising air near the equator, and a subtropical band of deserts.

As the power of computers increased, climate modeling groups began popping up around the world. By the mid- to late 1960s, weather prediction models were already quite accurate at forecasting the weather three days in advance, and the field of meteorology was entering a more mature, operational phase. Also, climate models came to stand squarely on the shoulders of weather prediction models. A good weather forecast was of tremendous importance to the economy, and as a result, the field of weather prediction began to receive more funding. There was a concerted push to improve the data being used to initialize the models. The use of spy satellites for weather "reconnaissance" had been proposed as early as 1950. By 1960, the Department of Defense had used classified spy satellite technology to launch the first weather satellite. By 1969, the design of the Nimbus-3 satellite proved helpful in improving weather forecasts. The satellite's infrared (IR) detectors could measure the temperature of the atmosphere at various heights all over the world. Ironically, if we remember that Richardson was a pacifist, the science of weather prediction was benefiting from money and technology that originated in the military.

Even with the ongoing improvements in weather data and computer technology, the emerging field of climatology was struggling to avoid the old adage about computers, "Garbage in, garbage out." When trying to represent global climate, scientists encountered a mind-bogglingly complex system. This was an enormous intellectual challenge. In addition to an atmosphere, climate models include land surfaces, oceans, sea ice, and hydrology—variables that made climatology much more difficult for the primitive computers, and for the scientists crunching the numbers. Also, the climate models needed to run for a much longer time, since instead of dealing with the few days needed for a weather forecast, the scientists were trying to simulate over decades, centuries, and in some cases even thousands of years. These scientists were tackling an immense problem. They were, in a sense, building the Earth from scratch. But along the way they were coming to understand important differences between predicting the weather one day ahead and predicting the climate 100 years ahead.

As Richardson learned the hard way, good data are important. And in his case, not having the precise starting point or initial conditions of the atmosphere took an otherwise great weather forecast and put it on the road to ruin. This dependence on initial conditions showed just how valuable good data are: they enable useful forecasts to go out a week instead of only a few days. Through advances in technology, scientists were able to enhance their data and thus design more accurate weather forecasts, creating predictions that were more accurate and less vulnerable to variation than ever before.[2]

Interestingly, climate models and weather models are often one and the same. But while climate models simulate actual weather, their results are analyzed differently from weather models. Climate prediction is not nearly as dependent on initial conditions as weather prediction is. In other words, the climate at the end of this century won't care very much about the weather at the beginning of this century. Climate is not nearly as chaotic as weather (for example, we can easily predict that July will be hotter than January). Climate model output is often analyzed by studying the season-to-season, year-to-year, and even decade-to-decade evolution of the climate. Unlike weather forecast models, they never attempt to predict precisely what a single day will look like. Instead, they look at how the statistics of weather change.

This is a very important distinction between weather and climate models: for climate forecasts, the initial conditions in the atmosphere are not as important as the external forcings that have the ability to alter the character and types of weather (i.e., the statistics or what scientists would call the "distribution" of the weather) that make up the climate.

These forcings include, for instance, the Earth's distance from the sun; how many trees are growing on the surface of the Earth; and, of course, how much carbon dioxide is in the atmosphere. You can't use models to simulate changes in the climate unless you know what will happen to the forcings.

And then there are the actual equations that make up the model. Climate models are built from two types of equations. First, there is the physics, which comes in the form of elegant equations such as Newton's laws of motion and conservation of energy. Second, there are equations, known as parameterizations, that are derived from observations and attempt to represent our current understanding of certain aspects of climate and weather. The physics in these models is universal, whereas parameterizations can vary depending on the team building the model. Parameterizations are a way to estimate all the complicated interactions that have been observed in nature but whose physics can't be directly represented in models due to limitations in computer resources and speeds. Each model uses different parameterizations to approximate what it cannot represent directly. As a result, different models predict different degrees of warming.

Because parameterizations inevitably introduce uncertainty, climate assessments typically draw on the collective wisdom of about twenty climate model projections, making up an *ensemble* of model simulations. This ensemble approach gives a better estimate of reality than any one particular model (though some models are better than others). Choosing the ensemble average is a way of drawing on multiple models to reach a consensus, rather than relying on any single model. Weather forecasters do the same. The assumption is that the approximation errors among models tend to cancel each other when we average their projections. As a result, the common, most robust tendencies are captured.

As computational speed and observational data continued to increase and improve, climate model simulations began to look more and more like the real world. It was eventually clear that climate models were ready for prime time; they were good enough to work on the problem of global warming. Those grueling calculations that Arrhenius had labored over could now be done quickly and rather painlessly by computers.

———

In general, there are two types of climate model runs that test the impact of global warming on the climate system: *transient* runs and *equilibrium* runs. In a transient run, greenhouse gases are slowly added to the climate system and the model simulates the impact of the additional $CO_2$ at each time step. In an equilibrium run, the atmospheric $CO_2$ level is instantly doubled, and the model is run with the higher $CO_2$ level until the climate has fully adjusted to the forcings and has reached a new equilibrium. The global average change in surface temperature due to the doubling of $CO_2$ is a number referred to as *climate sensitivity*.

In 1967, Manabe's group carried out the first series of climate sensitivity experiments using a very simple equilibrium model that represented the atmosphere averaged over the entire globe. The goal was to estimate what the Earth's average temperature would be if the level of $CO_2$ in the atmosphere doubled. This was similar to what Arrhenius had done by hand in the 1890s when he estimated that the planet would warm about 8°F. Using his one-dimensional model, Manabe came up with a different number: about 3°F to 4°F. Later, in 1975, Manabe and his collaborator Richard Wetherald published an analysis using a more advanced model that they had designed. This time they came up with roughly 6°F. Though this number was also less than what Arrhenius had come up with, it was taken much more seriously, since it had been derived from methods and a model that were more rigorous than the earlier attempts.

By the end of the 1980s scientists were also working on transient runs of the climate system, testing their climate models with varying levels of $CO_2$ to see what the future might look like. And even when these climate models were still in their infancy, they pointed toward an interesting result. When oceans were included in this model world, they acted to delay the appearance of global warming in the atmosphere for a few decades. They did this by soaking up some of the extra heat. Some people see this time lag as a gift, in the sense that it allows us an opportunity to prepare for and adapt to the coming climate changes. But many see the time lag as a curse, because it gives us reasons to procrastinate.

And it's tempting to procrastinate if you don't trust the models. But in the 1980s, climate models were beginning to show a very interesting consistency. You could start twenty different models with twenty different initial conditions, but the runs would all converge when they estimated the change in average annual global temperatures. They would, of course, show random variations in weather patterns for a given region or season, but every single model got steadily warmer over time.

The problem with verification of such results is that it's not possible to jump to the end of the century to see if a climate model is any good. But scientists can get around this by using their models to simulate events that have already happened. This simulation is called hind-casting, and it's an efficient way to test whether a climate model is skillful. Successful hind-casting experiments boost our confidence that climate models can capture past events and therefore can serve as a decent guide to the future. By successfully hind-casting a number of past situations (the effects of volcanic eruptions, seasonal variations, etc.), we can increase confidence in model simulations of the future. Basically, we can't prove that the models are right until the future happens, but we can prove that the models function by using certain rigorous tests.

Scientists have performed hind-casting studies on several major events in climate history to test how well the models can reproduce the climate at those times. They've modeled the height of the last ice age about 20,000 years ago, known as the Last Glacial Maximum (LGM), as well as a regional cooling event in Europe and North America roughly 500 years ago, known as the Little Ice Age.

There are also a few, rare opportunities to run a climate model in *forecast mode.* In June 1991, the eruption of Mount Pinatubo in the Philippines provided a perfect natural climate experiment. Pinatubo had injected about 20 million tons of sulfate aerosols into the stratosphere and created the largest cloud of volcanic aerosol haze and the largest perturbation to the stratospheric aerosol layer since the eruption of Krakatau in 1883. The haze spread around the Earth in about three weeks and attained global coverage after about one year.

Jim Hansen, a leading climate scientist at NASA's Goddard Institute for Space Studies (GISS) in New York, recognized this as a great opportunity to perform a real-time experiment: to use a climate model to predict how the real world would respond before it actually responded. In other words, his team would use the model to make a climate forecast that could be proved correct or incorrect in a relatively short time. So Hansen and his team added the Mount Pinatubo eruption as a forcing to the GISS climate model and made a prediction about how much the planet would cool over the coming year: about 1°F globally. They also predicted that the cooling would be concentrated in the northern hemisphere and would last about a year. The test involved waiting to see how skillfully the model had captured the real-world cooling. In 1992 there was a pause in the long-term warming, much to the delight of those who were skeptical about global warming. The average global temperature dropped roughly 0.9°F. Roughly a year later, the cooling began to subside and the steady uptick in global temperature resumed. The results were in, and the climate models were proved correct.

Since Manabe's first experiment with doubled $CO_2$, equilibrium runs have been performed thousands of times using increasingly sophisticated models. Climate models have reached a level of maturity approaching, if not rivaling, that of weather models. Whereas Manabe's 1967 model was simply one big grid square meant to cover the entire planet, today's climate models have more than 1 million grid squares that cover the planet. Each grid square is about 70 miles by 70 miles, with twenty-six vertical layers in the atmosphere. The next generation of models will resolve down to 30 miles by 30 miles. And as computers get faster, the resolution will improve further. It's not impossible that models will one day be able to predict the climate for every square mile on the planet.

Computers have already become a lot faster. In the 1970s, a century of climate took more than a month to run. In the current version of the National Center for Atmospheric Research (NCAR) T85 model, a century's worth of climate takes as little as thirteen days. Keep in mind that these new models have not only smaller grid boxes but also much more realism, which requires doing more calculations in less time. Perhaps far more telling, despite the impressive advances in data collection, modeling, and computational strength, climate sensitivity hasn't changed very much. The climate sensitivity estimated by the top global climate models ranges from 3.6°F to 8.1°F for an atmosphere going from about 300 to 600 parts per million (ppm) of $CO_2$. This is not far different from Manabe's estimate of 6°F in 1975 or Arrhenius's calculation of 8°F in 1896. It raises the question: how many more times do we have to do this experiment before we believe the answer?

Beyond these specific temperature increases, climate models help us see that global warming isn't just what's going on at the poles; these models also reinforce trends that you yourself have probably noticed in your lifetime. In the United States, spring now arrives an average of ten days to two weeks earlier than it did twenty years ago. Many migratory bird species are arriving earlier. For example, a study of northeastern birds that migrate long distances found that birds wintering in the southern United States now arrive back in the Northeast an average of thirteen days earlier than they did during the first half of the last century. Snow cover is melting earlier. Plants are blooming almost two weeks earlier in spring. The ranges of many species in the United States have shifted northward and upward in elevation. For example, a study of Edith's checkerspot butterfly showed that 40 percent of the populations below 2,400 feet have disappeared, despite the availability of sufficient food and shelter. These are all further reflections of the warming taking place right now.

And then there are the results from climate models. Climate models help us to understand what is happening and why. The experiment itself is fairly straightforward. You take the observed temperature record of the past century and compare it with the temperature simulated by a climate model driven by *natural events* such as volcanic eruptions and *human activities* such as combustion of coal, oil, and natural gas. Accounting for just natural factors, the models simulate the behavior of what is called the *undisturbed climate system* for periods as long as thousands of years—if external conditions like solar radiation remain within their normal bounds for the whole period. In other words, the model simulates the climate of an Earth without us, an Earth undisturbed by burning fossil fuels and by deforestation.

When you take us out of the calculations, you take out all the greenhouse gas emissions human activities have caused since the industrial revolution provided fuel for our cars and factories and large expanses of forests were cleared for agriculture and development. The rationale is simple. If a climate model, run with only natural forcings, cannot re-create the strong warming since the 1970s, then the real world is currently doing something Mother Nature cannot do on her own. If you can establish this, then you've successfully established that the temperature trend is truly exceptional.

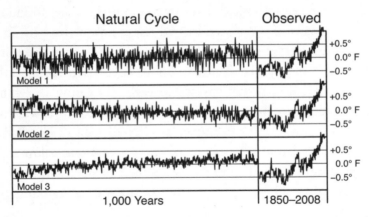

Natural variations in temperature of three different 1,000-year climate model simulations compared with observed data for 1850–2008. SOURCE: ADAPTED FROM STOUFFER ET AL., 1999, BY R. ZIEMLINSKI, CLIMATE CENTRAL.

The accompanying figure sums this up handily. It shows the natural variations in temperature of three different 1,000-year climate model simulations.[3] The variability in these different models is obvious: some periods are warmer, some cooler. But not one of these simulations captures any sign of an extended upward temperature trend. There isn't a single computer model simulation, called a *control run*, that exhibits a trend in global temperature as large or sustained as the observed temperature record. Hence, there is no way to explain the recent warming in terms of how the natural system has behaved over the last 1,000 years. If the recent warming trend were a result of natural forcings, then, assuming the models are correct, the model simulations would capture it and you would see a match between the observed record and the climate model. In fact, there isn't a single model that is able to produce a trend comparable to what we can see in the real world. Houston, we have a problem.

———

Climate models are not only important for showing us what could happen but are also a valuable tool for showing how much of it is our fault. With hind-casting, scientists can use climate models to isolate the physical fingerprint of human activity and figure out where the heightened levels of carbon in the atmosphere are coming from. Here's how it works. Different forcings—such as changes in solar radiation, volcano eruptions, or fluctuations in greenhouse gas concentrations—imprint different responses, or fingerprints, on the climate system. In the real world, these forcings are superimposed, one on top of another, making it difficult to assign blame to any single one. Therefore, climate models are used to make sense of the impact of each forcing, estimate the individual contribution of that forcing, and test whether it is responsible for the warming trend.

These forcings can be natural due to changes in solar radiation and volcanic eruptions, and they can be human-induced factors such as greenhouse gas concentrations. To repeat: climate models are used to calculate the fingerprint of each individual forcing, and thereby to distinguish how each forcing affects changes in temperature.

Take volcanoes. The idea that volcanoes affect climate has a long history. In 1784, Benjamin Franklin spoke of a constant dry fog all over Europe and North America that prevented the sun from doing its job and kept summer temperatures much chillier than usual. Franklin correctly attributed the dry fog to a large Icelandic volcano, called Laki, that erupted in 1783. In North America, the winter of 1784 was the longest and one of the coldest on record.

There was ice-skating in Charleston Harbor; a huge snowstorm hit the South; the Mississippi River froze at New Orleans; and there was ice in the Gulf of Mexico.

Scientists now know that volcanic eruptions, if large enough, can blast gas and dust into the lower stratosphere,[4] the layer of the atmosphere that begins about 6 miles above the Earth's surface. The strong winds at these altitudes, about 10 to 15 miles up, quickly disperse the volcanic material around the globe. The main gas emitted by the volcanoes, sulfur dioxide, eventually combines with oxygen and water to form sulfuric acid gas. This gas then condenses into fine droplets, or sulfate aerosols, that form a haze. The volcanic haze scatters some of the incoming sunlight back to space, and as a result temperature at the surface of the Earth, sometimes quite drastically, plummets for two to three years.

We've long known that solar radiation—like volcanoes—has the ability to affect global temperature, especially because the output of the sun is not constant. The sun has a well-established, roughly eleven-year cycle of *total solar irradiance*, during which its brightness changes over time. However, satellite measurements of total solar irradiance since 1979 show no increasing trend that could be responsible for global warming. The solar cycle is simply not strong enough to provide the temperature boost we have observed and measured. In addition, the eleven-year cycle is just that, a cycle—not a trend.[5] The sun is big and powerful, but its fingerprint simply does not match the observed warming. Its fingerprint is that of slight warming everywhere, including the stratosphere. If changes in solar output had been responsible for the recent climate warming, both the troposphere and the stratosphere would have warmed.[6]

No one is saying that solar variability and volcanic eruptions aren't important forms of climate forcing over the Earth's history. Climate model experiments show that the sun and volcanoes have indeed played an important role in changing temperature at time-scales ranging from decades to centuries. In fact, climate model experiments show that prior to the industrial era, much of the variation in average temperatures in the northern hemisphere can be explained either as episodic cooling caused by large volcanic eruptions or as changes in the sun's output.

The problem is that changes in solar output and new volcanic eruptions simply are not powerful enough to generate the large temperature rise we're currently witnessing. All of the testing finds that these natural factors cannot explain the warming of recent decades. Climate models can accurately estimate how much warming these natural factors produce, and—to repeat—they do not have the strength to generate the temperature increase we're seeing.

The only climate models that are able to simulate the changes in temperature we saw in the twentieth century are those that include natural forcings as well as human forcings—such as greenhouse gases.

Hind-casting isn't the only way to prove that the carbon we're producing is raising temperatures. Interestingly, scientists have learned that not all carbon in the air is the same; in fact, the carbon that comes from us bears our distinct fingerprint, a chemical smoking gun that shows just how much of this problem really belongs to us.

It turns out that just as humans come into this world with unique sets of fingerprints, so too does carbon. Carbon enters the atmosphere from a lot of different places, and each place stamps the molecules of carbon dioxide with unique fingerprints before sending them off into the atmosphere. Volcanoes emit $CO_2$ into the atmosphere when they erupt; the soil and oceans release $CO_2$ into the atmosphere; and plants and trees give off carbon dioxide when they are cut or burned. Burning coal, oil, and natural gas releases carbon into the atmosphere to form carbon dioxide. When you have the right tools, distinguishing where an individual molecule of $CO_2$ comes from is not hard.

Tracing carbon is a bit like tracing a bullet back to the gun it was shot from, and as with a ballistic test that links bullets to a gun, it helps to understand that not all carbon is the same. Carbon atoms (like many atoms) have variations known as isotopes, and these different isotopes are found in varying amounts around the atmosphere. Some got there from the oceans, others from volcanoes, and others from us. Carbon 12, 13, and 14 are all examples of carbon isotopes that are found in the atmosphere, and each comes from a different combination of sources. Each source, to repeat, has a unique chemical fingerprint. Carbon from the oceans, the atmosphere, and the land contains a healthy mix of carbon 12 and carbon 14. But carbon from fossil fuels has almost no carbon 14 at all.

Scientists use an instrument called a mass spectrometer to measure the amounts of carbon isotopes in the atmosphere and track the origin of the carbon. The mass spectrometer is very precise; it knows exactly which isotope of carbon it is measuring because the different carbon isotopes have different masses. So, the mass spectrometer can distinguish a carbon 12 atom from a carbon 13 atom from a carbon 14 atom. With a spectrometer, scientists can trace where the $CO_2$ in the atmosphere originated by measuring the ratios of the different carbon isotopes. In other words, a spectrometer can say whether a sample of $CO_2$ came from the ocean or from a volcano or from burning a fossil fuel.

According to precise measurements from mass spectrometers at several locations around the globe, the carbon dioxide molecules currently in the atmosphere have very little carbon 13 and carbon 14. Using this chemical signature to trace the carbon back to its source tells us that the increase we're seeing in atmospheric $CO_2$ did not originate in the oceans. Carbon dioxide that outgases from the oceans is *not* depleted in carbon 13. This also means that the increase in atmospheric $CO_2$ did not originate from living plants and animals, because $CO_2$ from organisms is *not* depleted in carbon 14. The chemical fingerprints of the extra carbon dioxide in the atmosphere match only the fingerprints of coal, oil, natural gas, and deforestation because these are the only sources that produce carbon dioxide depleted in carbon 13 and carbon 14.

It's true that most of the carbon dioxide in the atmosphere today comes from natural sources. But most of the *additional* $CO_2$ that's been placed in the atmosphere over the last 250 years comes from us. And it's the additional $CO_2$ that's raising temperatures. In terms of molecules of carbon dioxide, roughly one out of every four $CO_2$ molecules in the atmosphere today was put there by us.

All this carbon fingerprinting and the various climate models add up to one inescapable reality: the predictions that scientists have been making for the last twenty years have been getting more accurate. Weather forecasts started out as shaky, debatable calculations but evolved into a system of forecasting that virtually everyone in the world now relies on; similarly, climate prediction has evolved to a point where its results are sounder than ever before. No matter how many different ways the scientists run it, the results come out the same—a warmer planet that's getting warmer as a result of our carbon emissions.

There's no realistic way to take comfort in what these numbers are telling us. The forecasts that the models lay out is dire, and even though we don't see them every night on our local news, we cannot ignore it. So if climate models can show us that the temperature is rising and that it's our fault, when will our weather start to reflect our predictions about the climate? The short answer is that it already has.

# 4

# EXTREME WEATHER AUTOPSIES AND THE FORTY-YEAR FORECAST

I like watching basketball, but I'll admit that most of the time I can't keep up with it. Until the perfect moment when someone flies through the air and makes a basket, I can't see a damn thing. To me, the game is a lot of noise with very little signal. I'm much better off if I can watch a game with an aficionado. Watching the climate is no different. Learning how to see the climate system in play is like learning how to see the Lakers run a screen or watching a northeaster swing down from Canada. This is an art and a science, and it takes a trained ear to hear through the noise. Sometimes, hearing through the noise requires help in the form of a slow-motion instant replay. This is especially true when scientists attempt to understand the connection between global warming and weather that's happening to us right now.

The weather isn't what it used to be. In fact, all the data we've collected over the past fifty years point to the fact that the weather is getting more extreme.[1] But trying to isolate the fingerprint of global warming within the weather is much harder than isolating the fingerprint of global warming within the climate system. That doesn't mean it's not there; it just means that discerning climate change in the weather is a much noisier, more chaotic, and more complicated process. Ultimately, as in sports, the statistics can help us find the story buried beneath the noise. And climate scientists have come up with some very clever variations on using a slow-motion instant replay of the weather to help them understand how the statistics of extreme events are changing.

It turns out that you can use climate models as an instant replay to re-create a specific weather event. Think of this as like an autopsy, except that it's being performed on a specific extreme weather event. And although it can't determine the individual cause of the weather event, it can allow scientists to calculate the odds of such an event. Those odds can speak volumes if you know how to read them. The models can measure how much global warming shifted the odds in favor of allowing a specific type of weather event to happen. And, perhaps even more strikingly, the models allow us to see how those odds will play out and change in the future.

The European heat wave of 2003, an extreme weather event that killed more than 35,000 people, offers the best example of how climate models can help us see the global warming embedded within our weather. Public health officials were shocked at the scale of the human casualties caused by the heat. The largest number of casualties was in France, where almost 15,000 people perished in the first three weeks of August. Climate scientists were equally shocked at how far outside the range of historical temperature the heat wave registered. The summer of 2003 has been described as the biggest natural disaster in Europe on record.

The heat wave was dramatic. Temperatures in France soared to 104°F and remained unusually high for two weeks. There were extensive forest fires in Portugal, burning an estimated 1,500 square miles. Melting glaciers in the Alps caused avalanches and flash floods in Switzerland.

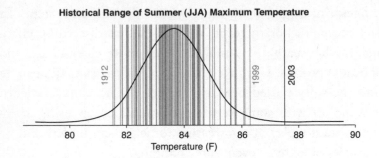

Each vertical line represents the mean summer temperature for a single year for the European region that spans [10W to 40E, 30N to 50N] over the period 1901 through 2003. Extreme values from the years 1912,1999, and 2003 are identified. SOURCE: ADAPTED FROM SCHÄR ET AL., 2004, BY C. TEBALDI, CLIMATE CENTRAL.

When we step back and compare the summer of 2003 with past summers, the picture becomes even more obvious. As you can see in the accompanying figure, there are a series of vertical lines that look rather like a bar code.[2] Each vertical line represents the mean summer temperature for a single year from the average of a region over Europe that spans [10W to 40E, 30N to 50N] over the period 1901 through 2003. Until the summer of 2003, the years 1912 and 1999 stood out at the edges as the most extreme temperatures in terms of hot and cold summers. Climate scientists estimate that the summer of 2003 was probably the hottest in Europe since at least A.D. 1500.

If climate is what you expect and weather is what you get, then the summer of 2003 was far outside what anyone would have expected. Statistically, in a natural climate system with no man-made $CO_2$ emissions, the chance of getting a summer as hot as 2003 would have been about once every thousand years, or one in 1,000.

The point of this weather autopsy isn't so much that the 2003 heat wave was, or was not, caused solely by global warming. Indeed, almost any weather event can occur on its own by chance in an unmodified climate. But using the climate models, it is possible to work out how much human activities may have *increased* the risk of the occurrence of such a heat wave. It's like smoking and lung cancer. People who don't smoke can still get the disease, but smoking one pack of cigarettes a day for twenty years increases your risk of developing lung cancer twentyfold. Thanks to some sophisticated climate models and well-honed statistical techniques, scientists can identify the *push* that global warming is giving the weather.

Step one was to re-create the 2003 heat wave in a high-resolution climate model using data observed during the heat wave. Scientists set up two sets of climate model experiments. One simulation included human-induced greenhouse gas emissions; the other simulation didn't include them. Like other climate model experiments, this one essentially created two worlds: a world with human influences and a world without them. By comparing the two, the scientists could look at what the risk of having a very hot summer is now and compare that with what the risk would have been if there hadn't been any human-influenced climate change. The difference between these two sets of odds would tell them just how much of a role humans played.

This weather autopsy, published in the science journal *Nature*,[3] showed that human influences had at least doubled the very rare chance of summers as hot as the one Europe experienced in 2003. The climate models showed that greenhouse gas emissions had contributed to an increase in such summers, from one in 1,000 years to at least one in 500 years and possibly one in 250 years.

What is perhaps most shocking is what happens when we run the models in *forecast* mode instead of autopsy mode. If the summer of 2003 had been a freak of nature, we could just chalk it up to chance. But the latest climate models paint a very bleak picture. According to their predictions, by the 2040s such summers will be happening every other year. And by the end of this century, people will look back wistfully to 2003 as a time when summers were much colder. Hindsight, as they say, is twenty-twenty.

In the United States, average annual temperature has risen more than 2°F during the past fifty years, and the temperature will continue to rise, depending on the amount of heat-trapping gases we emit globally.[4] Along with the general increase in average annual temperature, most of North America is experiencing more unusually hot days and nights, as well as heat waves.[5] Across the United States, ever since the record hot year of 1998, six of the last ten years (1998–2007) have had annual average temperatures that have made the record books, and these six years have ranked in the hottest 10 percent of all years since 1895 (when record keeping started). Also, the United States has seen fewer extremely cold days during the last few decades. In fact, the last ten years have brought fewer severe cold snaps than any other ten-year period since records began to be kept in 1895. There has been a decrease in frost days and a lengthening of the frost-free season over the past century as well.

By 2050, mid-range emissions scenarios predict that a day so hot that it is currently experienced only once every twenty years would occur every three years over much of the continental United States. By the end of the century, such a day would occur every other year, or more often. As for the cold, we'll be seeing even less of it. By 2100, the number of frost days averaged across North America is projected to decrease by one month; and decreases of more than two months are projected in some places.

The extreme weather of climate change is not limited to heat waves; the climate models suggest that other forms of extreme weather are also expected to increase. A warmer climate increases evaporation of water from land and oceans, and it allows more moisture to be held in the atmosphere. In other words, as the air gets warmer, it can hold more water vapor. Coupled with other warming-related changes, this additional moisture-holding capacity increases evaporation and will result in longer and more severe droughts in some areas and more flooding in others.

These trends are already beginning to be seen in the United States, and depending on where you are, you may have experienced them yourself. In the Northeast, the Midwest, and Alaska, the additional atmospheric moisture has contributed to more overall precipitation.

In the West and Southwest, the opposite is happening, as those areas have seen reductions in precipitation and increases in drought.

A warmer climate also means that it rains harder when it does rain. Diagnostic analyses have shown that with higher temperatures, a greater proportion of total precipitation comes from heavy precipitation events, such as blizzards and rainstorms. Heavy precipitation events averaged over North America have increased during the past fifty years, keeping pace with the increases in atmospheric water vapor that come from higher man-made carbon emissions.

Rain is just the start. The extreme weather produced by a warmer climate might include hurricanes. Because global warming results in warmer oceans, most experts on hurricanes agree that this warming of the ocean waters will make hurricanes more powerful. Hurricanes derive their energy from the evaporation of seawater, and water vapor evaporates more easily when it's warmer. The data suggest an increase in the number of more intense hurricanes in the North Atlantic during the past few decades. But more data are needed before scientists can be certain. Most climate models do predict that the strongest tropical cyclones will get stronger as global warming continues, but some models suggest that the total number of storms may actually decrease. There are several factors that influence the formation of tropical cyclones, including wind patterns, ocean currents, and local weather conditions. Any one of these factors might change in a warming world, in ways scientists are not yet able to predict.

But even if the jury is still out regarding the specific future of hurricanes, everyone is certain that damage will get worse. This will be due partly to population increases along the coast, and partly to the fact that global warming melts the ice caps at the poles, thereby raising the sea level. With a higher sea level come higher storm surges and more damage to our coastlines.

———

Ultimately, these extreme weather autopsies confirm something that many of us have long suspected: the weather is getting more extreme. The conditions have arisen for more major storms, longer droughts, and serious flooding, and they are getting worse.

These predictions and our seeming inability to heed their warming is a potential tragedy, reminiscent of Greek tragedy. Climate scientists seem to have become Cassandras, though the questions remain whether our forecasts will be heeded, and whether the harrowing scenarios of life on a warmer planet will come to pass.

The latter question is what Part II of this book is about. It explores the Achilles' heel of seven locations around the world, looking at how climate change will remind each one of its own frustrating vulnerabilities time after time if we humans continue to emit carbon at our current rates. Part II is about the unique perspective that each location offers on the risks of climate change to humankind. It's about using all the predictive tools that we've discussed—climate models, weather models, hind-casting, and extreme weather autopsies—to discuss the climate and weather scenarios for these various locations if global warming is allowed to continue unabated.

Just as the landscape of Earth is diverse and complex, so are the stories that specific landscapes will tell as climate change takes hold. These stories are the heart of Part II. Each chapter in Part II consists of two sections.

In the first section of each chapter, I've used climate models, environmental data, and—most important—the brightest scientific minds studying the climate in each location to help calculate and describe the specific risks each place will encounter over the coming decades. Despite all that math, models, and physics can tell us, these scientists are our most valuable tool for understanding the specific risks associated with each location. Because the climate of a place is so closely intertwined with the people who live there, it's nearly impossible to make an accurate prediction about the future without first understanding its most important initial condition: the human population. Along with all their measurements and readings, it is this understanding of the local population that these scientists strive for on a daily basis; the people's stories and experiences, what they've witnessed firsthand as they look climate change in the face every day, are as much a part of the predictions as the model-driven data. It is only by understanding the people who call each place home that we'll be able to predict how they'll react when climate change exposes their home to ever-increasing risks.

The second section of each chapter in Part II contains a series of predictions, based on a collection of climate models, which, taken together, offer a window into what the next forty years will look like for the place discussed in the chapter. The forecasts are works of fiction. I can't say whether the future will play out as I've described, but what I can say is that these predictions are based on the best available science derived from some of the cutting-edge climate models and real weather events from the historical record. In writing them, I have pored over reams of data with the aim of turning those numbers into stories, stories that give new visual guides to what climate change might look like in a place near you.

Each forecast begins with a glimpse of the regional climate during January and July based on two possible emissions scenarios. The scenarios are drawn from an average of at least fifteen different climate models, and each scenario makes different assumptions about future human activity—including greenhouse gas pollution, land-use alteration, technological development, and future economic development.

The first scenario is based on medium-high emissions. This scenario projects continuous population growth and uneven economic and technological growth. In it, the income gap between currently industrialized and developing parts of the world does not narrow. Heat-trapping emissions increase through the twenty-first century, and atmospheric $CO_2$ concentration approximately triples, relative to preindustrial levels, by 2100.

The second scenario is based on lower emissions. It characterizes a world with high economic growth and a global population that peaks by mid-century and then declines. There is a rapid shift toward less fossil fuel–intensive industries and the introduction of clean and resource-efficient technologies. Heat-trapping emissions peak at about mid-century and then decline. Atmospheric $CO_2$ concentration approximately doubles, relative to preindustrial levels, by 2100.

Some of these predictions have geopolitical implications; others have simply national ramifications. But one thing that's certain is that none of these scenarios will be happening in a vacuum. In a climate-changed Earth, every inch of land, ocean, and air will be affected. These stories represent some of the most dramatic and vulnerable locations, but they also represent places where the human species has been living for millennia, including some that will be rendered inhospitable by the changing climate. With regard to climate change, the worst-case scenario can be prevented through infrastructure investment (adaptation) and the adoption of clean energy technology and emissions reductions (mitigation). Some of these predictions examine what might happen if one location decides to adapt; others examine what might happen if a location refuses to change.

Ultimately, the basic assumptions driving these forecasts are that we will continue to burn fossil fuels, that the global population will continue to grow, and that because of both factors, greenhouse gases will continue to rise. The end result of this rise is that weather will not only become worse—it will become downright awful. Indeed, as the coming pages will show, the weather of the future and the way that weather affects life on Earth will be far worse than anything we've seen before.

Though these prophecies, as I've suggested, contain the seeds of a Greek tragedy, ultimately the forecasts also contain a kernel of hope, because unlike the prophecies in Greek tragedy, they are changeable. The forecasts paint a picture of just one *possible* future. While these forecasts, or indeed any forecasts, make certain assumptions about how trends will continue, one true variable they cannot approximate with much accuracy is our own behavior. We are the factor that could render all these predictions false, because we alone have the power to reduce our global carbon footprint. The question that we must now answer is how. In the end, these forecasts pose a question that is vital to our collective future: if we are really capable of forecasting the future and seeing the devastation of a changing climate in advance, will we act to prevent it? Can we rally around this forty-year forecast for the good of the world, or will we wait until the levees break before we decide to act?

# PART II

# THE WEATHER OF THE FUTURE

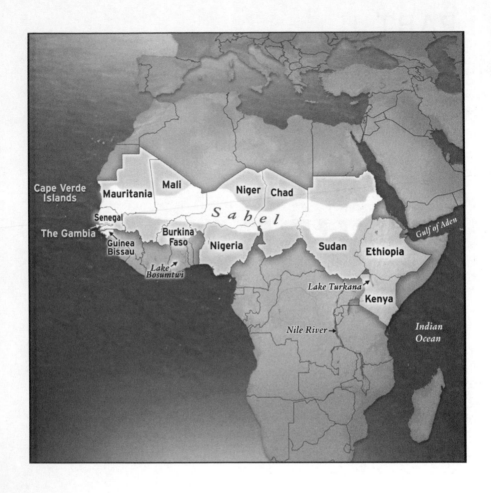

Cape Verde
Islands

Mauritania          Mali          Niger   Chad

Senegal                        *S a h e l*

The Gambia

Guinea          Burkina
Bissau          Faso          Nigeria          Sudan          Ethiopia

*Lake
Bosumtwi*                        *Lake Turkana*

Kenya

*Nile River*

*Gulf of Aden*

*Indian
Ocean*

# 5

# THE SAHEL, AFRICA

The word *sahel* comes from the Arabic *sahil*, meaning *shore*, and like so many things about the Sahel region of Africa, it is ironic. This is because the Sahel is a parched shore that both unites and divides the Sahara desert to the north and central Africa's tropical rain forests to the south. This pairing of opposites is a recurring theme in the Sahel, home to nomads and farmers, Muslims and Christians, Arabs and Africans.

The Sahel is also a place where the past and the future are sharply defined by climate. A semiarid savanna stretching out over 2,400 miles from the Atlantic Ocean in the west to the Red Sea in the east, the Sahel is consistently identified as one of the most vulnerable places in the world to global warming. But the Sahel, perched just on the southern fringe of the Sahara desert, is no stranger to hostile and recurring extremes in climate. You might say the climate history of the Sahel is simply a battle between trees and sand, greens and tans, wet and dry. Climate scientists have tracked this battle over millions of years using evidence left behind on land and in the sea. If you could transport yourself back 10,000 years, you'd see that this was a time when trees, not sand, dominated the landscape of the Sahara. Evidence pulled from the bottom of the ocean indicates that the climate of the Sahara was not dry but overflowing with tropical grasslands and forests and dotted with large permanent lakes—some as large as the United Kingdom, such as Megalake, Chad.[1]

But eventually green gave way to tan, and trees were once again replaced by sand. Today the Sahara is the world's largest warm desert. The Sahel has also seen significant change, slipping frequently into periods of drought. A 3,000-year climate record pulled from the mud at the bottom of Lake Bosumtwi in Ghana (see the map) shows evidence of *mega-droughts* lasting for centuries.[2] These shifts in climate have done far more than transform forest into desert; they've also altered the course of human history[3]—which is where this story about the Sahel begins.

Human history starts sometime around 6 million or 7 million years ago, when at least one species of tree-dwelling ape left its forested habitat in Africa and became the first member of the human family, the Hominidae. The species that paleontologists call *Sahelanthropus tchadensis*, recently discovered in central Chad, is thought to be the "last common ancestor" linking humans to chimps. At some point, an adventurous hominid rose up on two legs, and eventually hominids evolved into an early genus that paleontologists call *Australopithecus*. More recently, they evolved into our genus, *Homo*. This matter is still an active debate, but many scientists believe that the need to adapt to open grasslands helped shape the evolution of *Australopithecus*. Recent evidence suggests that the further evolution of these early hominids can, at least in part, be linked to the steplike shift toward cooler, drier, and more open conditions in the Sahel region.

The evidence for these steplike shifts in climate comes from the ocean floor. Long cores containing mud drilled from the bottom of the Gulf of Aden, off the coast of east Africa, contain almost 10 million years of climate history.[4] These cores contain three thick layers of dust, which originated in the Sahel and was picked up and carried by the northeast trade winds and then eventually dumped into the ocean. These thick dust layers represent drier conditions over the Sahel, and they appear at roughly 2.8 million, 1.7 million, and 1 million years ago, which happen to be very important junctures in human history.

The first shift toward colder, drier conditions (roughly 2.8 million years ago) marked a definite transition of the climate of the Sahel from dense woodlands into more open grasslands. As a result, many species saw their forested home replaced by a vast expanse of savanna. Some plant and animal species probably shrank in numbers as a result of this habitat loss, but many other species simply became extinct. The increased competition for food brought about by a changing landscape would have intensified the pressure to adapt.

During this time, at least two new hominid branches emerged. These two new branches in the early human family tree include the genus *Homo* and the genus *Paranthropus*. Paleontologists suspect that *Homo* was more of a jack-of-all-trades, whereas *Paranthropus* was more of a specialist and retained many common features of *Australopithecus*. As it turned out, *Homo* was far more capable of adapting to the changing environment and managed to come up with new strategies to cope with the increased competition for food.

For example, the first evidence of stone tools—crude choppers and scrapers—appears in the fossil record about 2.6 million years ago. The discovery of tool cut marks on mammal bones provides evidence of meat processing and marrow extraction. It is possible that *Homo* exploited different habitats by exploiting different types of foods. For example, eating meat and marrow would have marked an improvement over a strictly vegetarian diet because, unlike nuts and seeds, meat was available year-round. It is worth noting that the appearance of a bigger, more powerful brain in *Homo* coincides with the time of drying between 2 and 3 million years ago. Fossil records from the Turkana basin straddling northern Kenya and southern Ethiopia suggest that *Homo* species were characterized by a more delicate frame and smaller cheek teeth than their *Australopithecus* ancestors, and indeed had much bigger brains. In time this branch evolved into modern humans.

The other branch of hominids, the genus *Paranthropus*, with its very large teeth and a specialized chewing apparatus: a saggital crest on top of the skull that supported strong jaw muscles, tried to exploit a fading environmental niche, ate a mostly vegetarian diet and used those strong teeth and jaw muscles to crush nuts and seeds and grind the coarse vegetable matter that could still be found in the denser patches of savanna along rivers. This niche strategy is thought to have left *Paranthropus* more vulnerable, struggling to adapt to the cooler, more arid landscape, which provided fewer vegetarian options.

Some scientists suggest that during the next shift toward colder, drier conditions, about 1.8 to 1.6 million years ago, the branches of both *Homo* and *Paranthropus* underwent further splitting and pruning. At this time, the species named *Homo habilis* became extinct and our direct ancestor, *Homo erectus*, first appears in the fossil record. Fossil evidence suggests that by the third dry spell, about 1 million years ago, the entire *Paranthropus* line had become extinct and *Homo erectus* emerged as the winner, going on to occupy sites in North Africa, Europe, and western Asia—and ultimately evolving into *Homo sapiens*, modern humans.[5] The rest, as they say, is history.

———

Today, the Sahel is home to more than 60 million people. It is a place that has become synonymous with the word *drought*. In the past 100 years alone, the region experienced three devastating droughts. The first stretched from 1910 to 1916; the second stretched from 1941 to 1945; and then came the worst of all droughts, a long period of sustained declining rainfall beginning in the late 1960s and known simply as *the desiccation*.[6]

Climatologists estimate that from the 1950s through the 1980s, the Sahel saw rainfall decrease by about 40 percent.[7] This drought is linked to the deaths of more than 100,000 people, mostly young children; and it set off a wave of migration from north to south, from rural areas to cities, and from inland to the coast.[8] As a result, squatter settlements and urban overcrowding, accompanied by rising unemployment, increased. Political instability and unrest intensified across many countries in the Sahel. The dessication is considered by many scientists to be one of the most striking examples of climate variability the world has seen. A looming question now facing climate scientists is: when will another drought of this magnitude occur? Another question is: if and when such a drought recurs, who will emerge as the winner—the people or the sand?

During the 1970s, striking images of this crisis made their way out of Africa and reached television screens and magazine covers all over the world. The pictures of barren landscapes and children with haunting eyes and distended bellies led to coordinated international humanitarian efforts to help reduce the suffering. The crisis also revived a long-standing debate within the scientific community over the fundamental causes of drought. The debate centered on the concept of *desertification*, a process whereby productive land is transformed into desert as a result of human mismanagement.[9]

The issue of desertification dates back to the 1930s, during colonial rule in west Africa. There was a growing concern that the Sahara desert might be slowly creeping into the Sahel. The colonial regimes blamed desertification on the African people, specifically on rapid population growth and poor agricultural practices. It was a new twist to an old story. Instead of studying the impact of climate on human history, scientists were studying the impact of human history on climate. Man-made landscape alterations caused by overgrazing, intensive agriculture, and the cutting of trees were offered as a possible cause of the Sahel's drought.[10] At the time, some studies went so far as to suggest that at least one-third of the planet's deserts were a result of human misuse of the land.

In some respects, framing the problem as man-made allowed it to seem more readily fixable. With upward of 750,000 people in Mali, Niger, and Mauritania totally dependent on food aid and more than 900,000 people in Chad severely affected by the lack of rainfall, the west African countries of Burkina Faso, Cape Verde,

Guinea-Bissau, The Gambia, Senegal, Mali, Niger, Mauritania, and Chad became a formal geopolitical entity defined by a shared goal of combating drought. Formed in 1973, the Permanent Interstate Committee for Drought Control has a mandate to invest in research to ensure food security and to reduce the impact of drought and desertification.

Just as they had done in the past, the people of the Sahel were looking for ways to adapt and survive in a changing landscape.

If you look at rainfall records for the 1950s and early 1960s—before the drought began—the weather was actually a little wetter than average. This short-lived boost in rainfall allowed many farmers to grow crops in the northern Sahel, a region that was usually not suitable for agriculture. This type of rainfall-related migration is an age-old adaptation strategy for the people of the Sahel. But when the rains stopped, these crops were the first to go.

During and after the drought of the late 1960s, the only way to survive was to expand cultivated land. To do this, farmers had to cut down trees. By 1975, much of the remaining natural woodland had been converted to farm fields to feed a rapidly growing population. But by clearing native trees and shrubs, farmers were exposing their fields to the fierce Sahara winds; this exposure resulted in plummeting crop yields, and windblown sand buried entire villages. In what would have been an ultimate irony, it was suggested that these attempts to sustain life were actually what led to so many deaths.

———

Before you can begin to unravel the causes of drought in the Sahel and see how climate change will make those droughts worse, you need to understand why it rains there in the first place.

"That short answer is the African monsoon," says Alessandra Giannini. Giannini is a climatologist at the International Research Institute for Climate and Society (IRI) at Columbia University and has spent the past several years studying rainfall, or the lack thereof, in the Sahel. "Rainfall in the Sahel results from the collision between two different air masses; the moisture-laden southwesterly winds originating over the Atlantic Ocean and the dry northeasterly trade winds coming off the African continent."

In other words, if you're lucky and the conditions are just right, the collision of these two air masses will take place squarely over the Sahel and usher in welcome thunderstorms and much-needed rainfall.[11] But as history has shown, the Sahel doesn't always get lucky. The reason is that the Sahel sits at the northern edge of the African monsoon—and in some years the monsoon simply isn't strong enough to muscle its way that far north. When this happens, the northern part of the Sahel might as well be the Sahara.

Overall, the northern tier of the Sahel is rarely lucky with respect to rainfall; it averages only about 4 to 8 inches a year. To the south, the situation is a little better, and rainfall averages between 24 and 28 inches a year.[12] But even so, there is tremendous variability from year to year. Some climate scientists argue that the concept of average rainfall doesn't even apply in the Sahel. Also, the rainy season is short and intense—typically centered on August and lasting no more than four months. That means the dry season is very long, in a place where more than 80 percent of the people make their living growing crops and grazing livestock. June through September is known simply as the *hunger season*—the period when the harvest from the previous year has been exhausted and the next season's harvest is not yet ripe.

This is exactly why Giannini and her colleagues at the IRI have been working to develop seasonal rainfall forecasts for the Sahel. But Giannini knows all too well that before you can offer a forecast, you need to understand the past. And in the Sahel, that means understanding the causes of the drought.

"Two competing drought mechanisms were being floated around, one *local* and one *remote*," explains Giannini. "The local mechanism involved land use change on the ground in the Sahel. The remote mechanism involved changing the temperature of the oceans. But no one could really demonstrate which was the stronger influence." The local mechanism pointed the finger at human activity—for example, deforestation and overgrazing. The remote mechanism suggested natural causes. In other words, there were two options: the drought was made either by man or by Mother Nature.

Evidence that local activities can lead to local droughts was first proposed in the 1970s by Jule Charney, an atmospheric scientist at the Massachusetts Institute of Technology. The *Charney hypothesis*, as it came to be known, suggested that deforestation and overgrazing literally cool the land surface and ultimately decrease clouds and rainfall. Climate models were used to test this idea, and that is when the local mechanism began to unravel. Models looking solely at deforestation were unable to produce the kind of large-scale drought that was actually taking place in the Sahel.[13] Furthermore, satellite pictures of the Sahel confirmed that the land surface hadn't been changed nearly enough to alter rainfall patterns. Strike one for this hypothesis of human causation.

Giannini was interested in testing the other possibility—the *remote* mechanism. In the 1980s, a group of researchers from the United Kingdom's Meteorological Office had confirmed that changes in ocean temperature played a big role in generating the Sahel drought.[14] It was just a question of how big.

"I wanted to understand a very simple question," explains Giannini. "I wanted to see how well a climate model could reproduce observed rainfall over the Sahel." So she used an atmospheric climate model that took only one real-world factor into account: the ocean surface temperature. The model didn't know anything about the Sahel and its history of deforestation, desertification, and land degradation. As far as Giannini's climate model was concerned, no one even lived in the Sahel. And as it turned out, that didn't matter. "I couldn't believe it. The model reproduced the observed rainfall beautifully," explains Giannini.[15] "We all know these models aren't perfect, but the connection between ocean surface temperature and drought in the Sahel was very compelling." So much for blaming the drought on the farmers.

Giannini discovered that ocean temperatures helped regulate the strength of the African monsoon. What was even more fascinating was how each ocean played a specific role. On a year-to-year basis the Pacific Ocean has an effect on the Sahel's rainfall, thanks to El Niño. During an El Niño event, the Sahel is typically expected to experience a drought, whereas during a La Niña event, when the tropical Pacific is cooler, the African monsoon is expected to strengthen and rainfall is expected to be more abundant. The Atlantic Ocean and the Indian Ocean affect the Sahel's rainfall over much longer periods, from decade to decade. A warming of the Indian Ocean means a drier Sahel. And in the Atlantic, where the relationship is somewhat trickier, overall, when the southern hemisphere warms more than the northern hemisphere the rain belt across the Sahel is attracted farther south, toward the warmed hemisphere. This effect dries the Sahel.

"Ultimately, we found that you could explain the Sahel drought, as well as its persistence, by looking to the ocean temperatures," says Giannini.

Understanding this connection to the oceans provides the basis for seasonal rainfall forecasts for the Sahel. In essence, these seasonal forecasts are not attempting to predict how much it will rain on a specific day in the Sahel; rather, they predict when the rainy season might fail altogether or when large-scale flooding is likely. By knowing the present ocean surface temperatures, you can use climate models to forecast how the oceans will probably evolve during the next several months. Climate forecasts are an average of many climate models; similarly, the IRI uses numerous climate models with different conditions in the atmosphere to average seasonal patterns of temperature and rainfall. These averages give the most accurate predictions for the coming season's climate; indeed, seasonal rainfall forecasts for the Sahel have been issued since 1997, providing significant help in drought planning and food security.

But although the climate models rely on ocean surface temperatures to forecast rainfall and temperature, Giannini is quick to add that human activity does still influence the severity of drought in the Sahel. "If you cut down enough trees in the Sahel, there's no doubt it's going to get warmer and drier," Giannini explains. "What people are doing down on the ground can amplify the drought signal." And it is now well accepted that the combined effects of population growth, deforestation, overgrazing, and lack of coherent environmental policies, along with a significant decrease in rainfall, resulted in the crisis of the 1960s and 1970s, which was unlike any the world had seen before.

———

Of course, there is another, broader human influence that goes beyond the behavior of the local population. Global warming has already warmed up ocean surface temperature by about 1°F during the past century, and given the already established relationship between ocean temperatures and droughts in the Sahel, this warming trend will almost certainly have a negative impact on the amount of future precipitation there.

Global warming is affecting the region in ways that are not yet fully understood. If anyone can figure it out, Isaac Held would be a good possibility. Held is a senior research scientist at a division of the U.S. National Oceanic and Atmospheric Administration (NOAA): its Geophysical Fluid Dynamics Laboratory (GFDL), a prominent climate modeling center in Princeton, New Jersey. Few people understand the complexity of rainfall in the Sahel better than Held. A member of the Intergovernmental Panel on Climate Change (IPCC), Held served as a lead author of the IPCC's Fourth Assessment Report chapter on regional climate projections. The IPCC's regional projections use fourteen state-of-the-art climate models to provide a glimpse into the future. The GFDL climate model is one of the best in the world. And if you believe this model's projections for the Sahel, you'll be very worried about the future.

Held knows why the GFDL model behaves as it does; he just doesn't know if the real world will behave the same way. Remember that today drought in the Sahel is very sensitive to the gradient between ocean temperature in the north and the south. "The models all agree that if you warm the oceans of the southern hemisphere with respect to the ocean of the northern hemisphere, you dry the Sahel," explains Held. "But if you warm the oceans uniformly, there's no consensus among the models."

With regard to total seasonal rainfall in the Sahel, the models actually diverge—some predict more rainfall, and some predict less. Most of the models produce only modest changes out to 2100, but there were two outliers—one projecting a very wet future Sahel and one projecting a very dry future Sahel. The GFDL model was the dry outlier. It projected that summer rainfall in the Sahel would decrease 30 percent or more by the year 2100. Needless to say, a rainfall reduction at that level would be catastrophic. "Our model dries the Sahel in response to uniform warming. And that's why we dry so strongly into the future. It's the global warming signal that dominates," explains Held. "We're still trying to understand it." But in the meantime, Held points to model simulations that show a far more robust response.

The models all agree that the Sahel is going to get warmer. The IPCC estimates a warming of roughly 6°F to 10°F by the end of the twenty-first century.[16] A recent study looking at future heat extremes shows that heat waves will become longer and more frequent in the Sahel. And by the end of this century, the heat index across all of northern Africa will spike from May through October.[17] It is expected that the people in the Sahel region will be the most vulnerable, experiencing up to 160 days per year in the twenty-first century with a significant chance of heatstroke. Today, there are up to 180 days of medium risk in the Sahel and no high-risk days. By the end of the century, the people living in the Sahel are projected to experience the most severe increases in sunstroke in the world.

Regarding rainfall, Held also points to a recent set of experiments by Michela Biasutti and Adam Sobel of Columbia University that show a robust response among all the different IPCC climate models.[18] This particular study looked at the *timing* of the rainfall season rather than at the average amount of rainfall. "They see a delayed onset of the rainy season in almost all the models," says Held. In other words, the rainy season of the Sahel is projected to start later and become shorter, with storms that will possibly be more intense. This is not good news. (Case in point: on September 1, 2009, Ouagadougou, the capital of Burkina Faso, was hit by an unprecedented storm that brought more than 10 inches of rain in just a few hours. Widespread flooding left nearly 130,000 people homeless; they sought shelter in churches, mosques, and schools.)

Despite these dramatic images, and despite the fears about how global warming may affect the lives of those in the Sahel, Alessandra Giannini tries to remain hopeful.

"Honestly, I don't like to play this doom and gloom card with respect to the future. As climate scientists, we often spend our lives looking at problems from afar. But in the case of the Sahel, when you look closely at what is happening on the ground, you will be able to see pockets of resilience, pockets of adaptation." Those pockets of resilience are proof that people working together have the potential to overcome the forces of nature. Just ask Chris Reij, a scientist at the Vrije Universiteit (VU) in Amsterdam who specializes in soil conservation.

Chris Reij spends his life actively looking for ways to promote adaptation that will help the Sahel weather a climate-changed world. "I must admit that I'm doing everything possible to abandon research now. I've changed from soil conservation research to development action," Reij says. "We clearly have enough information to act."

The reason to act is that the pockets of resilience Giannini mentioned are, in fact, pockets of trees, millions of them.[19] Satellite images and tree inventories have found that the Sahel has become greener over the past thirty years.[20] Needless to say, the exact cause of this greening is still not perfectly resolved. Some scientists think the Sahel is simply bouncing back from the gradual improvement in rainfall since 1988; others think global warming could actually be helping to boost rainfall totals and spur on vegetation growth.

But Reij thinks the cause is the farmers. "You could say we've come full circle with respect to our ideas about drought and desertification," explains Reij. "Farmers were *not* the cause of the drought, but they are a big part of the solution." Reij has worked in the Sahel since 1978 and has never been one to spread doom and gloom. "I think there are a lot more success stories in the Sahel than we tend to assume," he says.

One success story involves farmers in Niger. A desperately poor country twice the size of Texas, Niger has seen about 12.4 million acres of trees, shrubs, and crops replace what was once barren ground.[21] Barren ground is all too common in Niger; four-fifths of the country sits within the Sahara desert. As a result, the vast majority of Niger's rapidly growing population is concentrated in the southern part of the country, the portion that sits in the Sahel. Niger has one of the highest population growth rates in the world: 3.3 percent, amounting to about 450,000 new mouths to feed every year. Niger's population has doubled in the last twenty years and each woman bears, on average, about seven children. If this growth rate continues, there will be 56 million people living in Niger by 2050. Experts have already begun to question how a country with a very small band of cultivable land can continue to feed itself, given this population growth and the looming threat of drought.

And yet, because of these conditions, an adaptation strategy is already in the works. In the long battle between trees and sand, the trees have begun to gain some ground—thanks to the help of local farmers.

"The story about the farmers is not a technical story. It's about a social process. It's about farmers who broke with tradition and villages that organized themselves," explains Reij.

The tradition of land "cleaning" and tree removal became common in the 1930s, when the French colonial government pushed Nigerian farmers to grow crops for export. Another by-product of colonialism was the fact that all trees in Niger had been regarded as the property of the state; this gave farmers little incentive to protect them. Government foresters were tasked with managing the trees, but oversight was lax and as a result trees were chopped down for firewood or for construction, without regard for the environmental costs. The loss of tree cover also led to a fuelwood crisis. Poor households were forced to burn animal dung or crop residues instead of using these for compost; that practice reinforced the downward spiral of soil quality and crop yields.

Despite these long-standing habits, in the mid-1980s Reij and his coworkers noticed a new trend among the farmers of Niger; they had begun to cultivate the trees that were on their property.

"When we asked farmers in Niger why they had begun protecting and managing their on-farm trees, the first answer was, 'because we must fight the Sahara,' " explains Reij. "And to them 'fighting the Sahara' meant fighting dust storms." Early in the rainy season the winds from the north tend to pick up. During the drought of the 1960s and 1970s, the winds would bring tremendous amounts of sand with them. As Reij puts it, "The sand acts like a razor. And it was just cutting down their young crops and carrying off the topsoil."

Roughly 85 percent of the almost 14 million people who live in Niger subsist on rain-fed agriculture, with millet and sorghum making up more than 90 percent of the typical villager's diet. "The farmers would have to replant three or four times before a crop would eventually succeed," says Reij. And so, in the mid-1980s, farmers decided to do things differently. Instead of clearing the land of trees, they started to protect their trees, meticulously plowing around them when planting millet, sorghum, peanuts, and beans. "It's the local farmers who are the real heroes here," says Reij.

By 2007, somewhere between one-quarter and half of Niger's farmers were involved in regreening efforts, and it is estimated that at least 4.5 million people had seen the quality of their lives improve significantly.[22] Over time, farmers began to regard the trees in their fields as their property. And in recent years, the government has come to recognize the benefits of this strategy and has allowed individual farmers to own trees. Farmers now make money by selling parts of the tree: branches for fuel; leaves; seeds; fruit for food. Over time, those sales generate far more income than simply chopping down the tree for firewood. As a result, the farmers protect this source of income. Crop harvests have also risen. With the trees come better diets, improved nutrition, higher incomes, and an increased capacity to cope with drought. Many rural producers have doubled or tripled their incomes. In some villages, the annual hunger season no longer exists.

Three factors play a role in regreening, or what Reij refers to as farmer-managed natural regeneration (FMNR). "First, despite the deforestation that took place in the 1970s, there was still a rootstock in the subsoil. And that rootstock was still alive," explains Reij. "So as the rains gradually returned, and the soil and the trees were protected, the trees began regenerating." The second factor is livestock. "Livestock grazes and digests the seeds. That means when the seeds pass through the intestines of the livestock, they will germinate more easily." And as it turns out, not all supposedly barren soil is actually barren. "There's a beautiful word for this," explains Reij; "it's called the *seed memory* of the soil." Under the right conditions, seeds that have been dormant for more than a decade will suddenly start sprouting again. "And there you have the beginning of your trees," Reij says, smiling.

When Reij talks about a new green revolution in the Sahel, he means it literally. But it's not so much about planting trees—an expensive proposition that had been attempted unsuccessfully in the past—as it is about recycling them. In Niger, farmers have protected and managed about 200 million new trees during the past twenty-five years. The number of trees that have actually been planted in that same period is only about 65 million. "But of the 65 million trees that have been replanted, at best 20 percent have survived, leaving only about 12 million planted trees," explains Reij. "So there is a lesson to be drawn from this; tree planting can help, but protecting and managing natural regeneration is much cheaper and produces quicker and better results."

Reij says, "The point is, you can solve the problem of climate change and the problem of poverty in parallel. That's the nice thing." In essence, the trees set off a chain reaction that improves the local economy as well as the environment. "With more tree species in the system, you increase biodiversity. And at the same time, it means you produce more fodder, which means you can sustain more livestock," explains Reij. Perhaps the biggest benefit has come to many of the poorest members of Nigerian society—women and young men. "It's a lot easier for women, because they can prune those trees and they have the firewood immediately at hand," explains Reij. And as for the young men, the annual exodus to search for higher-paying jobs in urban areas has slowed, thanks to new opportunities to earn income in an expanded and diversified rural economy. "So you have all kinds of positive spin-offs," says Reij. He points to some recent economic research, which suggests that increasing agricultural production by 10 percent can reduce rural poverty by somewhere between 6 and 9 percent.

The trees also act like a buffer during worst-case scenarios, and such scenarios may happen more frequently in a climate-changed future. In 2005, a year when the western Sahel saw flooding, the rains failed yet again in Niger. Reij visited villages with trees and villages without trees. He found that famine was much less of a problem in villages with many trees than in villages with few trees.

Likewise, villages with many trees did not suffer any drought-related infant mortality. "People told us that because they had more trees, they could cut and prune trees and sell firewood on the market. They could then use that money to buy cereal to feed their children," says Reij. "No doubt it was a brutal situation, but at least the children made it through." Reij knows this is not what perfection looks like, but for right now, it's close enough.

Ultimately, evidence from Niger demonstrates how relatively small changes in human behavior can transform the regional ecology, restore biodiversity, and increase agricultural productivity. Reij thinks such behavioral changes may even help bring rain back to the Sahel. "If you put a thermometer into barren, sandy soil you immediately get 120°F. But just 1 meter away, where you have some surface cover, the temperature immediately drops to 109°F," says Reij. "And with a bit of luck, if you have vast areas of regreening, the question is: might that begin to have positive impacts on local rainfall as well?" For Reij, this might just be the perfect answer.

In June 2009, Reij helped launch the Sahel Regreening Initiative in Burkina Faso and Mali. He says, "We thought: why not start an initiative which tries to build upon and scale up the existing successes in Niger? When we talk about adaptation to climate change, I am convinced that reforestation is a fairly effective answer. You improve the environment, you improve agricultural production, and you reduce poverty. There is every reason to be hopeful. We have the wind at our back."

The question is whether that will be enough. This is a matter of which forcing the Sahel will be most sensitive to in the coming decades—the oceans, the land surface, or global warming. The oceans have probably dominated the climate of the Sahel for all of human history. But as Reij has shown, the land surface is also clearly very important, especially in the Sahel, where it has the ability to outfit the landscape and help offset some of the changes that climate change will bring.

Unfortunately for the Sahel, Niger's tree experiment is an isolated event. Some recent climate model simulations propose that until 2025, the impacts of land degradation and vegetation loss over sub-Saharan Africa may be even more important than global warming for understanding climate change.[23] Until 2025, these models show that a drier, warmer climate goes with decreased agricultural production, on the order of 5 to 20 percent. Peanuts, beans, maize, rice, and sorghum will be likely to have the biggest drops in yield.[24] A recent study focused on Mali found a 17 percent reduction in crop yields out to 2040. Scientists consider these studies red flags. They recommend immediate action to develop more heat-resistant crops for the Sahel. In Benin, for example, a shift to yams and manioc is suggested as one adaptation strategy. And then, of course, there are the trees.

Reij points again to those pockets of resilience. "These predictions for 2050 and beyond, a 5°F increase in temp, a 20 percent decrease in crop yields—in the context of a doubling population, that's quite dramatic. That's why I look for villages where farmers were able to depress temperatures by 5°F using trees, and villages where the use of simple conservation techniques helped increase crop yields by 40 or 50 percent or even more. You have to be able to find examples that show it's possible to counterbalance what lies ahead." The people of the Sahel have a lot riding on this battle between trees and sand.

As scientists continue to search for perfect answers, red flags are clear on the horizon. Held has been watching as more red flags appear with each passing year and each new research experiment. And as he settles in to prepare the next round of IPCC simulations for the Sahel, he has one big concern of his own. "My biggest concern is that the GFDL model turns out to be right," says Held. If that happens, then sand may have finally won the battle in the Sahel once and for all. Unless, of course, like Reij, we abandon the search for perfect answers and simply begin to fight back. That might be the biggest Sahelian irony of all.

## Africa and the Sahel: The Forty-Year Forecast— Famine, Crop Loss, and Water Resources

| NIAMEY, NIGER | TODAY | | 2050 | | 2090 | |
|---|---|---|---|---|---|---|
| Emissions Scenario | JAN. | JULY | JAN. | JULY | JAN. | JULY |
| Higher | 75.8 | 85 | 78.5 | 88 | 82.5 | 92 |
| Lower | | | 78.5 | 87.4 | 80 | 88.9 |

## Forecast
## July 2015

There are few secrets in this parched, hostile landscape, especially about climate. As the models projected, the African climate was changing—becoming hotter and drier.

The trees were waging an all-out war against the brutality of the sun, and they seemed to be winning. Tens of millions of acres of Sahelian farmland had been planted with trees, and every leaf was viewed as a symbol of hope for the future. Agricultural productivity had increased, and there was actually a surplus of fuelwood. In the drier regions of Niger and Burkina Faso, people had begun reclaiming abandoned fields and getting new grain harvests by investing in simple water-harvesting technologies. Farmers were doing everything in their power to adapt to climate change, and the trees were their shield against the increasingly harsh climate. Through support from the Bill and Melinda Gates Foundation and OXFAM America, the regreening initiative had spread across the Sahel.

## November 2022

But there was only so much the trees could do. Without significant infrastructure investment, farmers across the Sahel were alone in their battle against the climate. Across the Sahel, yields of peanuts, beans, maize, rice, and sorghum were beginning to fall. Those farmers able to obtain heat-resistant seeds from international aid groups fared better. And those who shifted to more drought-resistant crops, such as yams and manioc, actually managed to put some food on the table. But ultimately, without rainfall, it was no secret that the farmers, who had lovingly planted and tended these trees as if they were their children, would be forced to watch as they eventually fell victim to the inevitability that was climate change. Climate change was leaving more and more people with less and less.

This basic realization led to a growing resentment across the African continent, because it was also no secret that the rich had caused the problem while the poor of the world bore the brunt of the impact. The experts said it would only get worse, and their advice was simple—adopt aggressive emission reduction goals and take steps to help victims of climate change adapt. They said it was a matter of national security.

To return to 2010: As it turns out, half of all $CO_2$ emissions come from only about 10 percent of the world's population. And that 10 percent includes Saudi Arabian oil moguls and Chinese investment bankers, not just rich Americans—the operative word isn't *American*; it's *rich*. The atmosphere doesn't care whether you drive a Ferrari in Dubai or Shanghai or New York. All it sees is the $CO_2$. As a result, several experts have come up with ideas on how to even the $CO_2$ playing field—spread the $CO_2$ wealth, so to speak. One group of scientists recently suggested a Robin Hood idea that essentially takes emissions (by means of a cap and tax) from the rich and distributes them among the poor. These scientists see their plan as a way to help lift people out of poverty.[25]

After all, $CO_2$ is just another term for energy. The World Bank estimated that if people in the United States swapped their SUVs (about 40 million, total) for fuel-efficient compact cars, the change would free up space in the atmosphere for about 142 million tons of $CO_2$. If you could magically convert that $CO_2$ back into energy and give it to the poor, it would provide basic electricity to about 1.6 billion people. In essence, everyone in Africa could have lights and running water.[26] To repeat, it seemed like a good way to even the playing field—and to decrease resentment. But in the end, no one really wanted to give anything up for the sake of strangers.

In 2015, this basic inequity between rich and poor remained a serious problem. And climate change, described by national security experts as a *threat multiplier,*[27] was turning up the heat, literally. Conflict was spreading across the African continent like wildfire. You needed only look at the past few years for evidence. The instability in the Sahel—especially Darfur—showed how quickly disputes over access to water and food during times of drought became politicized. The climate problem magnified preexisting threats stemming from ethnic and religious conflicts. The former UN secretary-general, Ban Ki Moon, had made that very point in the *Washington Post* more than 15 years earlier:

> Almost invariably, we discuss Darfur in a convenient military and political shorthand—an ethnic conflict pitting Arab militias against black rebels and farmers. Look to its roots, though, and you discover a more complex dynamic. Amid the diverse social and political causes, the Darfur conflict began as an ecological crisis, arising at least in part from climate change.
>
> Two decades ago, the rains in southern Sudan began to fail. According to U.N. statistics, average precipitation has declined some 40 percent since the early 1980s. Scientists at first considered this to be an unfortunate quirk of nature. But subsequent investigation found that it coincided with a rise in temperatures of the Indian Ocean, disrupting seasonal monsoons. This suggests that the drying of sub-Saharan Africa derives, to some degree, from man-made global warming.[28]

Experts were warning that attempts by the United States to build a "hearts and minds" coalition against Islamist extremism were being undermined by climate. They listened carefully to the latest tape from Osama bin Laden, on which he railed once again about the inequities of global warming and $CO_2$ emissions.[29] Jobs for young men in North Africa, to take just one example, had been further reduced by warmer temperatures and declining rainfall, intensifying resentment and unrest.

Scientists had even modeled the connection between conflict and temperature, to prove the point.[30] The data were painfully clear: conflict increased in lockstep with temperature. And when you looked at conflict combined with climate model forecasts of future temperature trends, there was a roughly 50 percent increase in armed conflict—almost 400,000 additional battle deaths—by 2030. The need to reform the policies of African governments and foreign aid donors to deal with rising temperatures had become urgent.

## March 2030

By 2030, piracy had become an epidemic on the order of HIV-AIDS. The piracy industry popped up after Somalia's central government collapsed in 1991. With no patrols along the shoreline, commercial fishing fleets from around the world came to plunder Somalia's tuna-rich waters. Initially, the pirates stepped in as a vigilante response to that illegal commercial fishing. Armed Somali fishermen confronted the crews of illegal fishing boats and demanded that they pay a tax. These were acts of desperation by local fishermen who had lost their livelihood. The entire country was, in fact, on the brink of starvation. It was not uncommon for pirates to smile when they were picked up by navy ships—they knew that they would at least get three square meals a day. But over time, these desperate bands of fishermen grew into something bigger, more organized, and much more sinister.

As rising ocean temperatures, pollution, and overfishing gradually erased their livelihood, more and more fishermen traded in their nets for machine guns and were hijacking any vessel they could catch: a sailboat, a yacht, an oil tanker, or a food ship chartered by the United Nations. Desperate times, they said, called for desperate measures. They said they had no other choice.

Because of all the hijackings, the waters off Somalia's coast were now considered the most dangerous shipping lanes in the world. The United Nations agreed to put a maritime peacekeeping force in place to patrol the waters, but there was only so much it could do. The main victims of the pirates were the Somali people. Nearly three out of every four Somalis had come to depend on food donations in order to survive. But the pirates were routinely overtaking the UN peacekeeping vessels, and the attacks made it very hard for the United Nations to keep sending provisions. Somali people were starving because the boats couldn't get through.

As of March 2030, no food ships had set sail from Mombasa to Mogadishu in months—the voyage was simply too dangerous. Despite the efforts of the international aid community and peacekeeping troops, conflict across Africa had increased along with the temperature. It looked more and more as though Africa was heading toward the status of a failed continent.

## January 2050

After years of conflict, drought, and food shortages, Africa was finally able to capitalize on something it has in abundance— sunshine. The Desertec project had been on the table for years, but two key elements were always missing: European funding and the support of African countries. When increasing demand from China and India sent oil and gas prices into the stratosphere, the Germans were eventually able to pull the money together, on the promise that Desertec would offset Germany's dependence on Russian gas supplies. The Desertec consortium, brought together by Munich Re, the world's biggest reinsurer, consisted of some of Germany's biggest and most powerful companies, including Siemens and Deutsche Bank. The plant symbolized a way to solve the problems of climate change and energy security simultaneously—and if everything went according to plan, it might help Africans as well.

North African governments sold their desert in return for water. Let us use your desert to generate power, the Desertec consortium argued, and you can use our energy to desalinate seawater so as to irrigate crops that will help feed your growing populations. North Africa's demand for water had, in fact, increased by two-thirds—an amount that was far beyond the available supply. For Africa, energy security was far less of a concern than water security. The deal was straightforward—Desertec would generate electricity for export in return for desalinating seawater for Africa—and it was a deal that was very hard to pass up. So with little to lose, the North African countries signed on, making one request—the plant was to be re-named Desertec-Africa.

The science was there. Every year, each square yard of the Sahara desert receives more heat from the sun than would be obtained by burning two barrels of oil. The calculations showed that all of Europe's electricity could be made in an area just 150 miles across. The Desertec-Africa plant used a technology known as concentrated solar power (CSP). The sun's rays would be concentrated by the use of mirrors, to create heat. The heat would then be used to produce steam to drive steam turbines and electricity generators. The advantage a CSP plant like Desertec-Africa had over standard solar photovoltaic panels, which convert sunlight directly to electricity, was that it had heat storage tanks. The tanks were able to store heat during the day and then power steam turbines during overcast periods, bad weather, at night, or when there was a spike in demand.

The cost overruns were substantial, and sandstorms and warring factions in North Africa made construction of the Desertec-Africa project an ordeal. But thanks to an important innovation that allowed the plant to use less water, construction was kicked into high gear. Like a standard coal- or oil-fired power plant, a solar thermal station requires large amounts of cooling water—something that is nearly impossible to come by in the Sahara desert. The water condenses the steam after it goes through the generator's turbines. But the innovation allowed the Desertec-Africa plant to be fitted with an air-cooling system that cut water demand by up to 90 percent— a huge break for investors. While many still argued that Desertec-Africa would make Europe's energy supply a hostage to a politically unstable region and that Europe was unfairly exploiting Africa for its sunlight, the project went ahead.

By 2050, Desertec-Africa was producing about half of Europe's electricity, with a peak output of 400 gigawatts—roughly equivalent to the output of 400 coal-fired power stations. The electricity generated by Desertec-Africa reached Europe via high-voltage power lines and trans-Mediterranean links that went from Morocco to Spain across the Strait of Gibraltar; from Tunisia to Italy; from Libya to Greece; from Egypt to Turkey, via Cyprus; and from Algeria to France, via the Balearic Islands. Part of a wider European super-grid that conveyed power generated from wind turbines in the North Sea, hydroelectric dams in Scandinavia, geothermal activity in Iceland, and biofuels in eastern Europe, Desertec-Africa had helped reduce European emissions of $CO_2$ by about 80 percent. The consortium had hopes of expanding. After all, a patch less than 450 miles across the Sahara could meet the entire world's electricity needs.

GREAT BARRIER REEF

South Pacific Ocean

Magnetic Island

Townsville

Queensland

# 6

# THE GREAT BARRIER REEF, AUSTRALIA

Joanie Kleypas and I first met in 2001, when I moved to Boulder, Colorado. I had just completed my postdoc at Columbia University and had decided to head west to begin work as a research scientist at the NCAR. Kleypas, a marine ecologist and geologist who uses climate models to study the health of coral reefs, had an office down the hall from mine.

At the time, I was studying a drought that had devastated a large stretch of central and southwest Asia. More than 60 million people across the region were in dire need of rain; and Afghanistan had been especially hard hit, as the drought came after two decades of political instability and economic isolation. I was looking at how large-scale climate patterns, like El Niño, might potentially help predict big droughts in the future and prevent such devastation. Whenever I needed a break, I'd stop in at Joanie's office to talk about her research on coral reefs. I remember thinking that Joanie was really lucky, because she studied beautiful places that were untouched by human activity. I loved listening to her talk about coral reefs and her travel to exotic places like Tahiti, the Caribbean, and, perhaps best of all, Australia's Great Barrier Reef.

I checked in with Kleypas recently to see how things were going with the Great Barrier Reef as well as to get a sense of what kind of impact global warming was expected to have on reefs in the coming years. She wasn't feeling so lucky. "I work on coral reefs, for God's sake. The entire coral science community is depressed," Kleypas admitted.

The Great Barrier Reef (GBR) is the largest tropical coral reef system in the world. It contains nearly 3,000 reefs built by more than 360 species of hard coral, and it attracts a wide variety of marine life. The GBR provides shelter to more than 400 types of sponges; 1,500 unique species of fish; 4,000 varieties of mollusks; 500 kinds of seaweed; and 800 types of echinoderms, which include starfish and sea urchins. It is bursting with activity, an oceanic metropolis analogous to Paris or New York. I grew up in New York, but given the choice, I would have much preferred growing up with a view of the GBR. It extends more than 1,200 miles along the northeast coast of Australia, and its area is about that of 70 million football fields, or half the size of Texas, which is where Kleypas is originally from.

As an undergraduate, Kleypas studied oceanography and marine biology at Lamar University in Beaumont, Texas. She had learned how to scuba-dive, thanks to her brother, and she dreamed of leaving the murky waters of the Gulf of Mexico for the majesty of the GBR. That dream came true when she received a Fulbright scholarship to study in Australia. She completed her PhD dissertation on the GBR and finally had a chance to see many of the things she only read about in textbooks. The GBR, home to six of the world's seven species of marine turtle, 30 percent of the world's soft coral varieties, and several hundred types of seabirds, became her backyard when she moved to Townsville, a small city of about 100,000 in the state of Queensland.

Townsville sits along the central part of the GBR and is a mecca for coral reef research. It's the site of James Cook University, where Kleypas studied, as well as the Australian Institute of Marine Science and the Great Barrier Reef Marine Park Authority. Townsville is about as close as you can get to the GBR and still breathe air. It is on the edge of the GBR lagoon, the reef just a thirty-minute ferry ride away. Kleypas did the reverse ferry commute for a while when she moved to Magnetic Island, population 2,107. More than half of this 20-square-mile island, known as "Maggie" to the locals, is designated a national park, so chances are you'll see more wallabies and koalas there than you will people. Kleypas rented a room from Lyndon DeVantier, one of the world's best field taxonomists for corals. "I would take my bike on the ferry and ride to the university in Townsville every day. It was an idyllic existence and I would have loved to have stayed," she says a little wistfully.

Kleypas began her career as a geologist interested in using corals to study changes in sea level. When she arrived at James Cook University, she and a team of researchers would go out for weeks at a time to remote sections of the GBR. They would cart out a small rig that allowed them to drill into the massive coral reef structures and extract cores. "Coral reefs are like big dipsticks," Kleypas explains. "They can tell us a lot about the geologic history of reef growth as sea level goes up and down." Despite being seasick from the strong currents and the 10- to 20-foot tides that swept the reef twice a day, Kleypas loved the GBR. And her hard work led to new insights about it. She and her team learned that since the end of the last ice age, the GBR had flourished or faltered depending on a delicate balance of conditions that included sea level, light, sediments, temperature, and circulation patterns. At the time, they weren't even thinking about carbon dioxide. They are now.

Corals have probably existed on the GBR for more than 25 million years.[1] The corals first formed during the geological era known as the Miocene. It was during the Miocene that India slammed into Asia and created the Himalayas. The Miocene also was a time when the Australian continent was on the move. As Australia tectonically made its way into the tropics, the shift to warmer ocean temperatures initiated the growth of some corals. Think of them as the very first version of the GBR, although back then the corals didn't form large structured reefs.

According to the Great Barrier Reef Marine Park Authority

(GBRMPA), the earliest record of complete reef structures dates back about 600,000 years. Research suggests that the current reef structure started growing above this older platform about 20,000 years ago, during the Last Glacial Maximum (LGM), the peak of the last ice age. At the time, much of the Earth's water was locked up in the form of ice, so the sea level was about 390 feet lower than it is today.

As the ice age came to an end, global temperature began to increase and the ice slowly retreated to the poles and mountain-tops where it had originated. By around 13,000 years ago, corals began to move into the hills of what had been Australia's coastal plain but was now underwater. At the time, the sea level was still 200 feet lower than it is today, but the coastal plain had already been swallowed up by the sea: only a few islands protruded out of the water. As the little islands slowly became submerged beneath the ocean, the corals finally had a place to set up shop in earnest. Scientists estimate that the present-day, living reef structure is between 6,000 and 8,000 years old; in other words, it dates from the period during which the sea level is thought to have finally stabilized.

Think of the modern GBR as a living veneer draped over ancient limestone. "The reef is like a layer cake," Kleypas explains; "every time the sea level rises, the reef adds a new layer." Today, that big layer cake helps to feed hundreds of thousands of Australians, because it brings in about $6.9 billion annually from tourism and other sources. What Captain James Cook identified as the perfect spot for a prison colony in 1770, when he surveyed the place then called New Holland, is now recognized as a World Heritage Site. So much for first impressions.

It's lucky for the GBR that Australia drifted into the tropics during the Miocene. Of all regions of the ocean, corals like the tropics best. Kleypas calls the reefs in this range, the band that stretches from 30°N to 30°S, the *vacation* reefs. The GBR alone has become a vacation destination for more than 2 million people each year.

You can find corals outside the tropics, too. A sturdy, brave few have inched their way out of their traditional comfort zone into places like Japan and Bermuda. But the farther poleward you go, the less the corals can build and actually reach the size associated with a true reef community. And it's the reefs that contain all the staggeringly beautiful biodiversity. "They get puny as you move north," Kleypas explains. "You can find single corals growing along the North Atlantic coast. But you'll never find a reef." The corals simply don't like cooler temperatures; coolness slows their growth rate, and in some cases colder waters can actually kill them. It seems corals are a bit like Goldilocks; they don't want things to be too hot or too cold.

———

By the time she left Australia in 1991 to begin her job at NCAR, Kleypas had fallen in love with the reef ecosystem that she had been using to reconstruct ancient sea levels. And so she shifted her focus from using the reefs as a tool to just studying the reefs themselves. And in the meantime, she's become a reluctant expert on how global warming is affecting these magnificent ecosystems.

"I really do think the coral reef community is suffering from some form of depression," Kleypas says. "It's like this. Imagine you fall in love with the most beautiful, amazing person. And then that person comes down with cancer. It's an incredibly sad thing. I fell in love with the reefs, not their disease." But it's the disease, the rising level of carbon dioxide in the atmosphere, that has captured the attention of an entire generation of marine scientists who are intent on saving the reefs. Kleypas is someone who uses climate models to study the reefs,[2] but, ironically, she could never have predicted that her own research career would come to this. She never expected to be forecasting the eventual decline of coral reefs.

That's not to say everything was perfect with corals before the impact of global warming became glaringly evident. Coral reefs were already showing signs of stress due to local-scale impacts such as agricultural runoff and destructive overfishing practices that include bottom trawling and dynamiting. The overall decline in water quality, due to pollution from coastal development, didn't help matters either. "Basically, the reefs are in worse shape the closer they are to people. The farther out you go, the better they look," Kleypas says.

But the global-scale stress due to climate change is adding a new dimension and a new threat to the overall resilience of the coral reef ecosystem. Global warming affects corals in two ways. The first is temperature: the oceans are warming up. The second is ocean chemistry: the oceans are also becoming more acidic.

Corals begin their lives as soft-bodied larvae that float through the water and eventually settle on a hard surface. As they settle, they also partner up with marine microalgae called *zooxanthellae*. Corals, which are animals, and their microscopic plant roommates are one of the prime examples of what scientists call a *symbiotic relationship*. Once the coral has partnered with the microalgae, it then sets to work building its skeleton by pulling dissolved calcium carbonate compounds out of the ocean water. The limestone skeleton forms the physical structure we think of as the reef.

A reef is the result of colonies of corals building their skeletons, like a bricklayer laying bricks, steadily over thousands of years. Shacking up with the algae turns out to be a big asset, as corals gain a second helping of food on top of what they are able to pull directly out of the water column. That second helping turns out to be very big. The zooxanthellae can provide up to 90 percent of the corals' energy requirements. "It's like if we had algae growing on our skin," Kleypas explains. "Whenever we'd go out into the sun, we'd get a jolt of extra energy courtesy of that algae." The corals get their energy from the plant by means of photosynthesis. This symbiotic relationship provides the extra boost that allows the corals to grow so large and form such elaborate reef structures. By definition, the symbiotic relationship benefits both partners. The microalgae get nutrients in the form of waste released by the coral. And of course it is the relationship with the algae that makes corals so pretty. The tissues of corals themselves are clear. Most of the beautiful colors of the coral reef, which can range from the palest pink to darkest black, are a gift from the zooxanthellae.

It's true that coral reefs like warmth. Ideally, they are adapted to water temperatures ranging between 65°F and 90°F. But corals don't like a quick spike in temperature. Go a little above the range they're used to, and trouble starts. That trouble comes in the form of coral bleaching.[3]

*Coral bleaching* is the term scientists use to describe the loss of all or some algae and pigment by the coral. As the algae are ejected, the white calcium carbonate skeleton becomes visible through the translucent tissue layer. The coral is weakened because it has lost the food energy provided by the algae. Already, in many places of the world—such as the Maldives, the Seychelles, and Palau—coral bleaching has effectively destroyed more than 50 percent of reefs. In the Caribbean, the numbers are worse, with between 80 and 90 percent of the reefs destroyed by bleaching, disease, hurricanes, and a number of problems related to coastal development, fishing, and other human activities. And future climate model projections, the kind that Kleypas works on, indicate that coral bleaching events are expected to become more frequent and severe over the coming decades.

"Bleaching is what happens when the coral kicks the algae out," Kleypas notes. "The way I have come to understand it is that the coral can usually handle the hot water, at least until you turn on the lights. When you add sunlight on top of a spike in ocean temperatures, photosynthesis kicks into overdrive." One by-product of all this photosynthesis is the release of too many free oxygen radicals. Free radicals are the same agents involved in the process of aging in humans. Crank up the photosynthesis, and you crank up the free radical count. "All those free radicals hurt the coral," Kleypas says.

The coral can't handle the imbalance, so it responds by evicting its tenant: kicking out the algae.

"It's almost like diarrhea," she says. "We have symbiotic microflora in our gut, and diarrhea is a common biological response when there is an imbalance." The diarrhea then serves as a way to get rid of things that are normally good for us. This classic tale of a good relationship gone bad makes you wonder if it might be a metaphor for the larger relationship between the planet and us.

Almost every reef region in the world has suffered extensive stress and in some cases death from coral bleaching. In 1998, a severe bleaching episode took out upward of 16 percent of coral reefs worldwide. El Niño played an obvious role in that event. During El Niño conditions, ocean temperatures in the Indian Ocean and in the central to eastern Pacific Ocean increase. Along with the warmer ocean waters comes a stable mass of high pressure, exactly the kind of weather pattern that ushers in a prolonged period of hot, sunny days. What might look like perfect weather is actually a condition for extensive coral bleaching. El Niño turns the lights on in a very big way.

"As a general rule, corals on the Great Barrier Reef start bleaching once temperatures exceed the average yearly maximum by about 2°F," Kleypas explains. In some cases it takes a much higher temperature spike to cause bleaching. But scientists say the range is generally between 2°F and 4°F. The actual temperature at which bleaching starts depends on how healthy a reef is to begin with, and how many other stressors, such as pollution and overfishing, are present. But as a general rule, if you crank up the temperature, you will most certainly crank up the stress level.

Average annual ocean temperature around the GBR varies from about 68°F to 79°F. A bleaching event is mostly likely to happen in January, February, and March—the southern hemisphere's summer months. Conversely, June, July, and August—winter in the southern hemisphere—are cooler. The all-important maximum temperature, which sets the baseline for when bleaching can happen, ranges from about 79°F at the southern end to 86°F at the northern end.

The fact that there is a range within which bleaching can happen means that other factors, such as global warming and just plain bad luck, are important in determining the ultimate fate of a coral reef under stress. Add a heat wave or an El Niño event on top of a week where ocean temperatures are at a maximum, and you've drastically increased the risk of a coral bleaching event. Global warming is no different. It effectively starts you off at a point closer to the temperature at which corals bleach. Water temperature along the GBR has increased by about 0.7°F since 1850, and the central and southern portions of the reef have warmed up even more, about 1.2°F.

The mass bleaching event of 1998 represents a turning point, according to many experts. They say that mass coral bleaching events have increased in extent and severity worldwide over the last decade. Prior to 1998, many reef systems had never experienced a severe bleaching event. But since 1998, every region has seen severe bleaching, and many regions have experienced significant die-offs because of warmer ocean waters.

And although the GBR is recognized as one of the best-managed and best-maintained coral reef parks in the world, it too has felt the effects of severe bleaching. The 1998 event hit 50 percent of the GBR. Another bleaching event occurred in 2002; this time, 60 percent of the GBR was hit. Fortunately, the death toll was low, but about 5 percent of the GBR is gone. In 2006, another bleaching episode hit the GBR, but it was more localized.

The southern hemisphere summer of 2009 appears to have been another bad break for the GBR. On March 16, 2009, the Australian government reported that a "weather triple whammy" had led to yet another round of coral bleaching. Stifling heat in December, floods in January and February, and winds from the tropical cyclone Hamish arrived one after the other. Ocean temperatures across most of the reef rose 3.6°F to 5.4°F degrees above the December average.

Russell Reichelt, chairman of the Great Barrier Reef Marine Park Authority, said that the "triple whammy" raised serious concerns about global warming. "The forecasts are an increased frequency of extreme events," he said. All these factors individually cause stress to the GBR. But, Reichelt added, it was their combined impact that was most worrying. "Historically, the reef has been resilient to events like this, but it is rare, possibly unprecedented, to have three such events in such a short period of time."

————

Coral reefs are complex and stunningly beautiful ecosystems. And their beauty attracts significant tourist dollars: it is estimated that tourism associated with the GBR contributes more than $5 billion to the Australian economy. This figure doesn't include other sources of revenue, such as commercial fishing. There are five main commercial fisheries operating in the GBR that together catch about 26,000 tons of seafood each year, with a total gross value of more than $220 million. But the reefs also provide benefits that don't have a price tag. Coral reefs not only serve to protect Australia's fragile coastlines from storm damage but also have been used to make several anticancer drugs. So it's quite accurate to say that coral reefs save lives.

Ultimately, extreme weather on top of the long-term global warming trend spells trouble for the reefs as well as for Australia's economy. Climate and economic models predict losses to the GBR tourism industry of between $95.5 million and $293.5 million by 2020, as a result of bleaching-related damage. And when you factor in the costs of all the other risks that warmer temperatures pose for Australia—extended droughts, heat waves, wildfires, and so on— you begin to sense that Australia is in grave danger.

Janice Lough, a researcher at the Australian Institute of Marine Science, explains the global warming effect as follows: "This seemingly modest increase in baseline temperatures has been sufficient to take corals over the bleaching threshold in 1998, 2002, and again in 2006. Modeling of future impacts suggest that a 1.8°F to 3.6°F warming of the GBR would result in about 80 to 100 percent bleaching compared to about 50 percent in 1998 and 2002." In other words, global warming makes bad luck worse. Lough adds that maintaining the hard coral at the heart of the reefs requires corals to increase their upper thermal tolerance limits by 0.2°F to 1.8°F per decade. But how do you teach corals to become more heat tolerant?

Bleached corals aren't dead; they're just starving. The loss of their energy-providing zooxanthellae means they're not getting enough food. If the stressful conditions come to an end soon enough—that is, if the weather changes and temperatures become cooler again—the algae can come back, and the corals can survive the bleaching event. But corals that do survive a bleaching event come out of it in a weakened state. As a result, they'll be likely to experience reduced growth rates, decreased reproductive capacity, and increased susceptibility to diseases. Because it can take up to twenty years for reefs to fully recover, these recurring bleaching events are a kick in the teeth. Prolonged bleaching often leads to coral death.

Complicating matters is the fact that saving the reef from severe bleaching events requires patiently nursing it back to health. The recovery process is time-consuming and requires recolonization by coral larvae. Even under ideal conditions, coral recovery is slow and may take decades. You need sufficient connectivity of *source* reefs—reefs that export fertilized coral and fish eggs to other reefs downstream—as well as good water quality to make sure that the spawning and recruitment of larvae will succeed. Bleaching will actually kill the corals if the stresses are too severe or too persistent. This situation is really not so different from a prolonged drought. The condition of a coral when it enters a bleaching event is likely to determine its ability to survive the bleaching event.

Corals get a good bit of attention because a bleaching event is highly visible and because so much money is tied up in the reefs. But other parts of the reef ecosystem are also vulnerable to temperature. Seabird chicks have undergone severe die-offs during periods of unusually high sea temperatures. These die-offs, called *nesting failures*, result when parent birds can't get dinner for their chicks. The fish they prey on follow productivity zones that are temperature-dependent. And when these fish change location, the birds can't always find them. Sea turtles are also at risk. The sex ratio of turtle hatchlings is temperature-dependent, and continued warming could cause a significant bias toward females in future populations.

But again, temperature is just half of it. The other half of the global warming situation is ocean acidification (OA). Kleypas describes OA as the "silent problem" associated with increasing $CO_2$. It's also been described as "the other $CO_2$ problem." The other $CO_2$ problem has scientists very worried. And, ironically, it comes as the result of a favor the oceans are doing us. No good deed goes unpunished.

Atmospheric $CO_2$ is currently at 387 ppm, but it would actually be a lot higher if not for the oceans. Roughly 30 percent of the excess carbon dioxide released into the atmosphere by human activities since the industrial revolution has been absorbed by the oceans. That's the favor. If not for the ocean uptake, atmospheric $CO_2$ would be on the order of 450 ppm today, a level that would have led to even greater climate change than is already under way. But this favor provided by the oceans doesn't come cheap. It has led to a roughly 30 percent increase in the concentration of hydrogen ions through the process of OA.[4]

This process, OA, is simply what happens when you add carbon dioxide to seawater. The additional $CO_2$ causes a slight reduction in ocean pH, which is a measure of how acidic or basic a substance is. The pH scale ranges from 0 to 14. Pure water, for example, has a pH of 7 and is considered neutral. A pH less than 7, as in vinegar and lemon juice, is acidic. A pH greater than 7, as in ammonia or laundry detergent, is basic. The ocean is also slightly basic; the average pH of surface seawater today is about 8.1. But there is reason to believe that this is not the number it should be or the number it will remain.

The United States is the third-largest consumer of seafood in the world, with total consumer spending for fish and shellfish at about $60 billion per year. Coastal and marine commercial fishing generates as much as $30 billion per year, and nearly 70,000 jobs. Healthy coral reefs are the foundation of many of these viable fisheries. Needless to say, there are plenty of reasons to be worried about OA.

Although it is described as "silent," OA is a straightforward consequence of rising atmospheric $CO_2$. This condition doesn't have a lot of the uncertainties that plague some other climate change forecasts. It's freshman chemistry. And ocean pH is something we're good at measuring. Since the 1980s, pH measurements collected in the North Pacific Ocean (near Hawaii) and in the Atlantic Ocean (near Bermuda) are registering a decrease in pH of approximately 0.02 unit per decade. Since preindustrial times, the average pH of ocean surface water has fallen by approximately 0.1 unit, from approximately 8.2 to 8.1, and it is expected to decrease further, depending on how high $CO_2$ rises. If atmospheric $CO_2$ concentrations reach 800 ppm, pH is predicted to rise an additional 0.3 to 0.4 pH unit.

What worries scientists is that even a slight decrease in pH does something funky to ocean chemistry, specifically to the amount of *carbonate ions*, a very important form of carbon. Corals pull in carbonate ions and secrete calcium carbonate ($CaCO_3$). This is a process called *calcification*, and it uses the dissolved carbonate ions to form calcium carbonate minerals for shells and skeletal components. Once dissolved in seawater, $CO_2$ gas reacts with the water to form carbonic acid ($H_2CO_3$), which can then break apart by giving up hydrogen ions to form bicarbonate ($HCO_3^-$) and carbonate ($CO_3^{2-}$). Increasing the amount of carbon dioxide dissolved in the oceans has a nasty side effect: it decreases the amount of carbonate ions in the water. Fewer carbonate ions means less material for building such things as calcium carbonate reefs and clamshells.

A study published in the journal *Science* in 2009 seems to confirm this.[5] Experts at the Australian Institute of Marine Science in Townsville looked at coral samples from the GBR over the past twenty years to track changes in growth rates. Specifically, they measured the rate at which corals absorb calcium from seawater to build limestone skeletons. The study concluded that corals in the GBR are growing more slowly. The team of researchers investigated 328 colonies of massive *porites* corals from sixty-nine reefs covering coastal as well as oceanic locations spanning the entire length of the GBR. Because the *porites* coral lays down annual growth bands, it's possible to count back to a specific year and correlate the growth during that year with the sea surface temperature over the same time period. Ten of the cores dated back to 1572.

The researchers sliced up the cores and used X-rays to measure three growth values: skeletal density, annual growth rate, and calcification rate. The values for growth and density allowed them to calculate annual calcification. They found that between 1900 and 1970 calcification rates increased 5.4 percent. But that's when you could argue that the party ended. Calcification rates dropped 14.2 percent from 1990 to 2005, mainly owing to a slowdown in growth. Researchers measured calcification as decreasing from 0.56 inch per year to 0.49 inch per year. Scientists can't confirm yet that what they are seeing is indeed the impact of increased OA, as opposed to other stressors such as coastal pollution. But the fact that the effect is seen on inshore as well as offshore reefs suggests to them that the cause is more likely to be *global* (for example, temperature and ocean acidification) than local (for example, pollution). It appears that not just the economy but also the GBR is in a recession.

And the impact of OA isn't limited to corals. "There's some very cool new research out there about clown fish," Kleypas says. "Of course, it's also very depressing." The clown fish has become an iconic species ever since Disney's blockbuster *Finding Nemo* gave kids a look at the biodiversity of the GBR. Interestingly, Phil Munday and his colleagues at James Cook University found that OA affects Nemo's ability to find his way home. "Baby clown fish use their sense of smell to find a suitable habitat. And ocean acidification impacts their ability to differentiate between what is a suitable habitat and what isn't," Kleypas says. Recent research suggests that OA has impaired their sense of smell. "The baby fish aren't getting the signal that says, 'Bad habitat; don't go there!' and are less able to sense the proper habitat. And the researchers were not subjecting the fish to huge changes in pH. They were consistent with future projections," Kleypas adds.

Research by Ken Caldeira, an ecologist, and his team at Stanford's Carnegie Institution for Science suggests that ocean pH has not been more than 0.6 unit lower than today's levels during any time over the past 300 million years. Yet the results obtained with the Stanford climate model show that the continued release of fossil fuel $CO_2$ into the atmosphere could cause an eventual pH reduction of 0.7 unit over the next 300 years.[6] An unprecedented change in pH over 300 million years is a lot easier to handle than the same change over three centuries. When $CO_2$ changes over a time interval longer than 1 million years, ocean chemistry is buffered by interactions with carbonate minerals, and that buffering helps reduce the impact of acidification. Caldeira's research suggests we're talking about such a severe mismatch in timescales that adaptation is almost impossible.

When I asked Kleypas what she envisioned the GBR might look like by 2050, she said, "The distribution of reefs will be more patchy. Biodiversity will go down. There will be more algae and less hard coral. Erosion will become more noticeable. There will be fewer baby corals. It's not all going to be dead; . . . the deeper parts of reefs may fare better."

Climate models support this grim snapshot of the future. The models suggest that if $CO_2$ emissions stay as they are, average water temperature in the GBR could increase by another 3.6°F to 5.4°F by 2100.[7] And studies have suggested that these increasing base-line temperatures, combined with the likelihood of more extreme weather events, like heat waves and flooding, could result in mass bleaching events every two to three years. Recent modeling studies indicate that if atmospheric $CO_2$ levels hit 600 ppm, it will be very tough to save the corals. By 650 ppm, it will be impossible to save them.

Still, Kleypas thinks the overall outlook for impacts associated with increasing temperature is a little rosier than the outlook for impacts associated with OA. "We know corals can handle high temperature. We see them in the Red Sea and the Arabian Gulf. Corals can get used to warmer water, the same way people living in Arizona and Phoenix don't suffer as much from heat stress." In places like the Red Sea and the Arabian Gulf, corals don't bleach until they reach temperatures about 18°F higher than their summer maxima, a much higher threshold than for similar species located in cooler regions. But there is one problem: the projected rate and magnitude of temperature increase will quickly outpace the conditions under which coral reefs have adapted and flourished during the past 500,000 years. Experts are worried that corals won't be able to adapt fast enough to keep pace with even the most conservative projections for climate change. "The problem is," Kleypas adds, "it takes time for corals to get used to the increased temperature." And, unfortunately, time is not on their side.

———

With regard to finding an approach to help the coral adapt, coral bleaching reveals itself as the kind of problem where traditional management approaches that focus on minimizing or eliminating sources of stress don't help much. The ocean is not like the heated pool at a hotel or motel: coral reef managers are constrained by a frustrating inability to directly turn down the ocean temperature when their reefs start to overheat. Needless to say, when you can't control the single most significant factor affecting a bleaching event, you're dealing with a challenging environmental management problem. To address the problem, in 2006 the Great Barrier Reef Marine Park Authority published *A Reef Manager's Guide to Coral Bleaching.*[8]

Resilient reefs seem to share a few important qualities. One is location: they are located in a zone of cooler water. Some sites may have consistently cooler water because of upwelling or proximity to deep water. A second quality is shade: some reefs may be protected from bleaching because their exposure to the sun is limited by topographic or bathymetric features. Reefs shaded by cliffs or mountainous shorelines may also have a reduced risk of bleaching. Many reef areas are unlikely to have features that can provide shade, but fringing reef complexes around steep-sided limestone or volcanic islands, as in Palau and the Philippines, may have many shaded sites. A third quality is screening. Naturally turbid conditions may filter or screen sunlight, providing a measure of protection for corals exposed to anomalously warm water. Ongoing research suggests that organic matter in turbid areas may absorb ultraviolet (UV) wavelengths and screen sunlight. Corals at these sites may be less susceptible to bleaching.

The goal is to establish a network of marine protected areas (MPAs). If you can identify MPAs or reef areas that are likely to be more resistant to mass bleaching, then these are the places that have the best shot at survival in a warmer world. And if scientists can help set up a network of resilient coral reef refuges, then they can draw on these like a garden to reseed coral reefs that have been hurt by bleaching. In the context of mass coral bleaching, these refuges can serve as seed banks or source reefs for less resilient areas. But if the special reef refuges are to serve this role, they need to be effectively monitored and protected from local stressors such as anchor damage, overfishing, and pollution.

One such spot that scientists and conservationists are working hard to protect is an area called the Coral Triangle, which spans eastern Indonesia, parts of Malaysia, the Philippines, Papua New Guinea, Timor Leste, and the Solomon Islands and contains 53 percent of the world's corals. It's often compared to the Amazon rain forest of Brazil because it has such a high level of biodiversity. The Coral Triangle covers an area of 2.3 million square miles, about half the size of the United States. It has more than 600 reef-building coral species—75 percent of all species known to scientists—and more than 3,000 species of reef fish. It also has the greatest extent of mangrove forest of any region in the world. Both the mangroves and the coral reefs serve to protect fragile coastlines from damage by storms and tsunamis. Scientists feel that if they can protect this reef system, it will help them save other reef systems.

More than 120 million people live within the Coral Triangle, and about 2.25 million depend on its marine resources for their livelihood. Between the tuna fisheries and the tourism, the estimated total annual value of the coral reefs is $2.3 billion.

The MPAs are carefully selected areas where human development and exploitation of natural resources are regulated to protect species and habitats. By providing refuges for exploited fish stocks, MPAs provide benefits for commercial fisheries as well. Healthy fish stocks in MPAs replenish surrounding fishing grounds with eggs, larvae, and adult fish. Right now only 1 percent of the ocean is protected, compared with about 12 percent of the land. The question is whether the existing MPAs are in the right places and where we should put the next ones. Kleypas explains that finer-resolution climate models can help scientists select the optimal locations for MPAs. Implementing this principle in MPA design involves considering prevailing currents and adjacent non-reef areas. Linking MPAs along prevailing currents that carry larvae can replenish downstream reefs, increasing the probability of recovery at multiple coral reef sites. Adjacent non-reef areas are important to connectivity because they can become important staging areas for coral recruits as they move between reefs and into new areas.

"I call this high-$CO_2$ window the Noah's ark period. We have to save as many species as we can," Kleypas says. "We also need to help make the reefs more resilient. The reefs that are in the best shape today are the reefs with the best management practices," she adds. "I'd like to see some advances in coral reef restoration and coral farming." In essence, this involves managing the ocean more as we do the land. It's interesting to imagine coral reef farmers growing and tending to baby coral reefs. And this may be the best hope we have.

Another important part of management is monitoring the reefs using satellites. Coral Reef Watch, a program of the U.S. National Oceanic and Atmospheric Administration (NOAA), has developed tools to analyze satellite images and help reef managers assess the likelihood of mass coral bleaching events. It's a little like a weather forecast for your coral reef, and it includes maps and indexes that track how warm conditions are getting in the tropics.

The maps use satellite data to show the intensity and duration of spikes in sea surface temperature. If you can monitor the intensity and duration of heat stress, you can get a sense of where a mass bleaching event might occur and how bad it might be. Both the intensity and the duration of heat stress are important factors in predicting the onset and severity of a mass bleaching event. The monitoring tool tracks the anomalous sea surface temperature, the difference between the observed ocean temperature and the highest temperature expected for a specific location, based on long-term monthly averages. It provides a useful reference point that shows the extent to which current temperatures vary from those that the corals are accustomed to experiencing at that time of year.

Temperature anomalies of 2°F to 4°F extending over a period of several days to several weeks should alert managers that there is a medium to high risk of bleaching. The NOAA Coral Reef Watch program also developed Tropical Ocean Coral Bleaching Indices to provide additional nearly real-time information for twenty-four reef locations worldwide.[9] For each reef site, the closest 30-mile satellite data are extracted, including present sea surface temperature, degree heating weeks, climatology, and surface winds. Visual warnings are provided for each site when conditions reach levels known to trigger bleaching in vulnerable coral species. It's a bit like the way forest rangers track the potential for wildfires on land.

There are also monitoring programs that encourage laypeople to serve as scientists. The sheer size and remoteness of many reef areas can be a substantial challenge for reef managers wishing to detect the onset of bleaching and monitor bleaching-related impacts. Reef users can help managers keep an eye on the reef during periods of high risk. A program in the GRB called BleachWatch engages people who love the reefs by teaching them how to help monitor coral bleaching events. BleachWatch provides an early warning system for coral bleaching and forms part of the Coral Bleaching Response Plan of the Great Barrier Reef Marine Park Authority's (GBRMPA). The program is aimed at tour guides and allows them to go about their everyday work, be it guiding snorkel trails or diving, while taking a mental picture of their "home reef." Back on the vessel, staff members fill in the monitoring form and send it back to the GBRMPA at no postage cost. In return for the monitors' efforts, the GBRMPA analyzes the information and provides monthly site reports.

Experts like Kleypas also hope that high-resolution climate models might help in planning for the Noah's ark period. The models can be used to fast-forward in time and get a better sense of what the reefs might look like as temperatures go up and pH goes down. The models might also be able to serve as a tool that allows a better understanding of which species are going to make it and which are more vulnerable. The species that are most resilient and most likely to survive can potentially be used to reseed reefs that have suffered from bleaching. Kleypas says that right now, most models look only at the surface ocean temperature. But when you begin to study the ocean at depth, the models could help identify places where the reefs are likely to survive. The models could actually help managers target specific areas to protect—areas like the Coral Triangle that will probably be used to help rebuild the reefs that are having a harder time.

Despite all these efforts, some people remain worried that corals simply will not have the resilience or the adaptive strength required to get past this high-$CO_2$ window. Such people are calling for more dramatic measures. Some have recommended setting up an underwater repository of corals similar to the Svalbard Seed Bank, a cave on the Norwegian island of Spitsbergen that (as its name implies) preserves thousands of plant seeds from around the world. Svalbard is, in a sense, an underground "doomsday vault" built to serve as the ultimate safety net for the world's seed collections, protecting them from a wide range of threats. In a repository, the corals would be saved from rising temperatures and OA, but Kleypas sees this as a last resort.

With all these tools and programs coming online, Kleypas is trying very hard to stay positive. As a scientist she is pragmatic, but as someone who is passionate about coral reefs, she conveys an uncommon sense of hope. "Scientists are very introverted people by nature. We don't tend to be inspirational. We make future predictions based on the here and now. But I've been trying to give people hope. I hate to give a doomsday lecture and tell folks the reefs are all going to hell. People don't know what to do with that information."

On the other hand, she doesn't want to come across as a Pollyanna. "We've already entered into this window of high $CO_2$. So, we have to aim for mitigation. We can't just stabilize emissions. We need to then get $CO_2$ levels down. I like to tell people we don't know how high the $CO_2$ is going to be, because that level is fundamentally up to us."

Kleypas is not one to shy away from the possibility of genetic engineering. "Scientists are thinking about the symbiotic algae," she explains. "The question is: can we seed a reef with algae that are more resilient to temperature changes? One theory that has emerged is that the amount of bleaching that occurs at each reef may be influenced to some extent by the prevalence of stress-tolerant algae. And so scientists have begun surveying corals for the presence of stress-tolerant zooxanthellae within reefs. There's still so much to learn about these symbiotic algae, but identifying which algae are most stress-tolerant may help managers to assess the potential resilience of different sites. We know how to engineer resilience in terrestrial species, but we know much less how to do it in marine ecosystems. And we need to figure it out," she says with a sense of urgency. "This is one of the most remarkable times to be a scientist. Sure, we can just sit back and watch it happen and confirm that our predictions are coming true. But that would be embarrassing." I suppose it's fair to say that we've reached a point where we need to make our own luck.

## The Great Barrier Reef, Australia:
## The Forty-Year Forecast—Coral Bleaching,
## Ocean Acidification, and Economic Struggle

| TOWNSVILLE, AUSTRALIA | TODAY | | 2050 | | 2090 | |
|---|---|---|---|---|---|---|
| Emissions Scenario | JAN. | JULY | JAN. | JULY | JAN. | JULY |
| Higher | 81.4 | 65.3 | 83.8 | 68.1 | 87 | 71.1 |
| Lower | | | 83.3 | 67.4 | 84.5 | 68.5 |

## Forecast
## March 2017

For several months, squadrons of scuba divers from all over the world had been heading out into the unusually warm waters off the coast of Townsville, tanks of air strapped to their backs and monitoring checklists dangling from their wrists. The divers were roaming the seas in search of bleached corals—a terrible job for anyone who loves the reefs. El Niño set off a worldwide coral bleaching event affecting hundreds to thousands of miles of reefs simultaneously. This El Niño came when ocean temperatures were already warmer than average and caused severe to extreme bleaching even along the very carefully managed and monitored GBR—with the result that more than half of the colonies turned completely white. The divers went out each morning to identify sick corals and came back each evening hoping that park officials wouldn't need to come up with a catastrophic level for coral bleaching, too.

The Coral Reef Watch program set up by NOAA had been monitoring ocean surface temperature using satellites and was able to provide scientists and volunteers with almost up-to-the-minute information. Temperatures along the GBR ranged from 80°F to 84°F, and bleaching was widespread. Reports from volunteer groups, such as BleachWatch and Reef Check, warned that large sections of coral were bleaching at levels much higher than those seen in 1998, when waters heated by El Niño killed 15 percent of reefs worldwide. In the GBR alone, reef damage related to bleaching had caused losses to the tourism industry on the order of $250 million; and we had felt sick to think that when all was said and done, we might lose more than one-third of these beautiful reefs.

It was hoped that local solutions—for example, marine sanctuaries and volunteer monitoring efforts—might create lasting changes. But it's tough to manage global warming locally. And in the end, the root cause of bleaching, warmer ocean temperatures, could be addressed only by a worldwide effort to reduce $CO_2$ emissions. Scientists had hoped that the more remote reefs, safely isolated from human impact, might fare better. They were wrong. The bleaching event caused extensive mortality in nearly every coral reef region in the world. No man is an island—and no reef is safe from the long arm of climate change.

## December 2019

As climate change continued to accelerate, the words *severe* and *extreme* were no longer adequate to describe the destructive potential of Australia's wildfires. That's why the residents of South Australia awoke on the first morning in December 2019 to face yet another sustained warning of catastrophic fire danger. The category *catastrophic* had been put in place by the Australian Bureau of Meteorology (BOM) in 2009, after the horrific wildfire called Black Saturday radically altered everyone's definition of how bad a wildfire could get.[10] The wildfires came on Saturday, February 7, 2009; and by the time the sun rose the next morning, they had killed 173 Australians and traumatized countless others. According to a report by the Victorian Bushfires Royal Commission,[11] the fires generated winds so strong "that trees appeared to have been screwed from the ground." At the time, no one dared imagine what Australia might look like if Black Saturday had gone on for a week. We found out in 2019. That catastrophic outbreak came to be known as Black December.

Fire marshals begged residents of the eastern Eyre Peninsula and the west coast districts in the state of South Australia to evacuate their homes immediately. If they had learned anything from Black Saturday—when many residents stayed on, hoping to defend their property but in the end losing their lives—it was that no one should attempt to be a hero. Fire authorities, however, had no official mandate, so they could not force people out of their homes; they could only beg the residents to leave.

It had been a brutal few months. In October, an ongoing drought had kicked up a thick wall of red dust that reduced visibility in Sydney to less than two city blocks. Snapshots taken of the Sydney Opera House—silent, ghostlike, and shrouded by a thick red veil—made their way around the globe, and the world looked on in fear and fascination. The ferocious wildfires, pervasive drought, and unbreathable air made Australia seem like hell on Earth. And for those who lived there it was.

In November, intense heat pushed the average temperature for the month into numbers never seen before. Towns throughout Victoria and southeast Australia were running 2°F to 4°F above the previous record, set in 2009. Melbourne was running 2°F above the record set during the previous November. The city's chief meteorologist summed it up: "Usually when you break records like these you break them by a tenth of a degree. But we're seeing we're two, three, or even four degrees above previous records. This is not natural." And by December, the heat and drought—together with low humidity and strong winds, created the perfect conditions for a catastrophic wildfire. Happy New Year, Australia.

## January 2025

After the widespread bleaching event of 2017, the idea of setting up a coral bank began to gain traction. The Svalbard Global Seed Bank had proved to be successful. Why not try something similar with corals?

The Smithsonian Institution in Washington, D.C., finally received funding to set up the Smithsonian Global Coral Vault. Corals from tropical oceans were being placed in deep freeze at the Smithsonian to preserve them for posterity as they faced destruction from rising greenhouse gas levels. This *coral cryobank* would ultimately house hundreds of samples from each species. The funding came after new research suggested that most coral reefs would be largely dead by 2040, wiped out by a combination of rising temperatures and increasing acidity in the world's oceans. The affected areas included Australia's 1,600-mile GBR, Caribbean reefs, and reefs in the Coral Triangle—an area spanning Indonesia, the Philippines, Malaysia, Papua New Guinea, and East Timor. Carbon dioxide emissions had risen above the safe level for corals, and reefs around the world were showing the impact. The Smithsonian's vault was a matter of reverting to plan B. And its very desperation re- flected the despair among scientists about rising $CO_2$ levels.

The coral vault applied a breakthrough deep-freeze technique developed by scientists to regenerate coral from frozen samples. The scientists took tiny biopsies from coral, froze them in liquid nitro- gen at $-330°F$, and then thawed them to regenerate polyps. These scientists were proposing to do the same for every species of coral on the planet. There are about 1,800 known tropical corals and another 3,350 cold-water species. The Smithsonian would house about 1,000 samples of each coral in a large room in a subbasement of its museum in Washington, D.C. The facility was nicknamed the "Morgue of the Sea."

## December 2050

The overall acidity of the ocean continued to increase. Corals reached the point where they were dissolving more quickly than they were growing. Consequently, many coral reefs were unsus- tainable. It was expected that pH levels would continue falling. By 2100, climate models forecast a further drop in pH of 0.3 to 0.5 unit—which would make the world's oceans more acidic than they had been in tens of millions of years. And while there might be thousands of coral polyps sitting in deep freeze at the Smithsonian, there was no ocean on the planet Earth that they could now call home.

# 7

# CENTRAL VALLEY, CALIFORNIA

At the end of the movie *Pretty Woman*, after Richard Gere has climbed up a fire escape and rescued Julia Roberts from the drudgery of the real world—we hear a baritone voice declare, "This is Hollywood, the land of dreams. Some dreams come true, some don't. But keep on dreamin'." It's a classic Hollywood ending. Everyone is happy and full of hope for the future.

Hollywood may serve as the unofficial capital of dreams, but it's certainly not the only place in the Golden State that lies within the realm of unreality. The Central Valley, where the Sacramento–San Joaquin Delta is located, represents a very different kind of dream for the future, a kind that many scientists have come to see as sheer fantasy.

In the Sacramento–San Joaquin Delta, the Sacramento and San Joaquin rivers converge into canals, levees, streambeds, marshes, and peat islands. With an area that spans about 24 miles east to west and 48 miles north to south, the Delta is the hub of California's water supply system. The entire state—especially the rapidly growing and increasingly dry metropolitan areas, such as Los Angeles and San Diego to the south—depends on this very small area of land for its water. The dream is that the Delta will be able to supply enough clean, fresh water to help cities and crops increase forever, all without harm to the natural environment.

If this sounds too good to be true, that's because it is. Those cities now stretch from San Francisco and Silicon Valley in the north to Los Angeles and Orange County in the south. And the crops, including alfalfa and corn, grow on a version of *Fantasy Island*. Many islands in the Delta are 15 to 20 feet below sea level and are protected only by increasingly fragile levees. The Delta was a good dream, a very American dream. And it even came true, for a while.

But today, the Delta has become an example of how complicated and costly it can be to sustain a dream, especially when global warming is a factor. California's gold rush attracted a new generation of Americans who came west and saw endless possibilities. But endless possibilities and unbounded growth require infrastructure. And infrastructure comes at a cost, both for the economy and for the environment.

I recently spoke with an economist and an environmental engineer about the future of the Sacramento–San Joaquin Delta. Ellen Hanak, director of research at the Public Policy Institute of California (PPIC), and Jay Lund, an environmental engineering professor at the University of California-Davis, are part of a multidisciplinary team that includes biologists, economists, engineers, and a geologist. The team members are studying the Delta and attempting to salvage what remains of the dream. I spoke with Hanak and Lund about climate change and about two reports they had recently co-written. The first report recognized that the Delta is in a crisis. The second report looked at ways to manage the Delta's water while protecting the Delta ecosystem. Management strategies that attempt to satisfy the competing interests of the Delta's economy and its environment have been discussed and debated for almost 100 years, and sometimes California's version of the battle between the North and the South has resulted.

Engineers are taught to find optimal solutions to problems. As applied to the Sacramento–San Joaquin Delta, *optimal* means finding a balance between environmental and economic interests, and between agricultural and urban users of water. *Optimal* doesn't mean perfect; it does mean that no one wins entirely. Fish will die and agricultural yields will decrease, but *optimal* means that the Delta will have a future. It also means writing a big check up front and making profound changes to the Delta landscape before Mother Nature or the climate makes its own changes for you.

"I'm not sure how it's going to play out. A lot of people are working to make sure nothing happens in the Delta, or that things only happen their way," says Lund.

The Central Valley stretches approximately 400 miles from north to south and is roughly the size of the state of Tennessee. Today, about 6.5 million people live in the Central Valley, which is considered the fastest-growing region in California. All the scientists I talked to when I was selecting the locations to write about in this book considered it a hot spot with regard to global warming.

The northern half is the Sacramento Valley, and the southern half is the San Joaquin Valley. The two halves meet at the Delta. California's capital, Sacramento, is located along the Sacramento River's banks. The rivers and their tributaries are harnessed by more than 100 large dams that produce the majority of California's hydropower. The Delta is part Frankenstein's monster, part natural wonder.

The Delta provides water for two out of three Californians, and for almost 4 million acres of farmland. At the Delta's western edge lies Suisun Marsh, and at its southern end are two prominent examples of California's water infrastructure: the Delta-Mendota Canal and the California Aqueduct. The canal and the aqueduct deliver water from upstream reservoirs and the melting mountain snowpack to cities and farms in every direction. The Delta is the hub of the state's water supply because it serves as the transit point for water. Whenever the Delta shows signs of its original personality and moves from being a predictable freshwater conveyance system toward being a vast tidal marsh, every effort is made to push it back into place, by repairing levees, releasing water from reservoirs, or reducing water exports. Needless to say, a lot of dreams rest on the earthen walls of the suboptimal canals and levees.

And that's the trouble with the Delta. Scientists say the canals and levees have become increasingly vulnerable to a catastrophic failure, whether it arrives abruptly in the form of an earthquake or slowly as the result of a rising sea level caused, in turn, by global warming. In any event, the scientists are nearly unanimous: the Delta is unsustainable.

At the start of the gold rush in the late 1840s, the levees provided a simple way to lock down the landscape and get more value out of the land. Lund, an expert on water resources in California, says, "If you look at the geological history of the Delta there was always a lot of variability." Native species, such as the delta smelt, came to depend on that variability. But there are now 1,300 miles of levees in the Delta and Suisun Marsh—they form a longer stretch than the entire California coastline—and the Delta's natural variability has been kept to an absolute minimum.

"The Delta is the direct result of rising sea level since the end of the last ice age," Lund explains. "During the last ice age, the Delta was to the west of the Golden Gate Bridge. But starting about 10,000 years ago, as sea level rose, the Delta moved inland. About 6,000 years ago, the Delta arrived at its current location, where, for the most part, it was able to keep up with sea level rise by building marshes. Sediment accumulation in the Delta kept up with the slow rise by forming thick deposits of peat." Peat is made of organic matter: decaying plants and animals that only partially decompose. This dead matter can't get enough oxygen to break down completely, because everything is waterlogged. "It took 6,000 years for that peat deposit to build, as one layer of new plant material grew on top of previous layers of peat," Lund says. Through this gradual process of flooding and rebuilding, a diverse, resilient ecosystem evolved. Then came the gold rush.

It was actually during the California gold rush that farmers stumbled on the Delta and struck their own kind of gold. The peat in the Delta was capable of producing excellent crops. But to farm the organic-rich soils, farmers first needed to drain the islands. After 6,000 years of continual flooding and rebuilding, the Delta was, for the first time, being pinned down. "This involved constructing levees around the islands, filling most tidal channels, and, most important, lowering local groundwater tables below crop root zones by constructing perimeter drains," Lund explains. "When you're located at the confluence of two major rivers, the dry period is when you want to grow your crops. And if you can keep the soil moist but not waterlogged all year long, then you've really struck gold."

The Delta was a perfect spot to settle and farm. Its proximity provided easy access to the miners and markets, and its soil was beyond comparison. "And so these natural levees were formalized and the islands were dried out," Lund explains. Today, the Delta grows more than 90 different crops, including wine grapes, pears, rice, corn, and tomatoes, producing more than $360 million annually in farm sales. But altering the landscape came at a price that has yet to be paid.

"The problem," Lund says, "is that peat soils are meant to be waterlogged. When they remain dry for long periods of time, they lose their integrity." So, every year you farm, you lose some soil. "Today, we're farming on borrowed time," Lund explains. "We're mining the soil until the islands fail." In just 150 years, about 6,000 years' worth of peat has been eroded. It's gone. By engineering the variability out of the system, we've attempted to pin something down that cannot be pinned down.

Pinning the Delta down also pushed it down, and so it is that much more vulnerable to earthquakes, flooding, and a rising sea level. When the levees were first constructed, no regulatory policies forced people to consider their impact on the Delta ecosystem. Farmers were doing their own engineering, and they were optimizing around just one quantity—agricultural yield. Hanak, the economist, explains, "Many of the water diversions upstream and within the Delta were made before the public demanded environmental protection."

Levees built 100 years ago confined water to channels and transformed the Delta from marshland into dry islands of land available for human use. The Delta islands started sinking when the marshlands, the source of all the fertile peat soil, were first drained. And the sinking—the scientific term is *subsidence*—continues to this day. There are now seventy-four islands in the Delta. Most of them are below sea level, many by more than 20 feet. Subsidence also increases seepage into the islands, raises the likelihood of levee failures, and increases the costs and consequences of catastrophic island flooding. Take your pick: earthquakes, floods, a rising sea level, subsidence, and urbanization all contribute to the increasing likelihood of multiple levee failures. Scientists, not known for hyperbole, describe this as a catastrophic failure.

———

The Delta has far more in common with New Orleans than with Hollywood. You could argue that when the Delta experienced a major levee break in June 2004, cracks in the dream were also beginning to appear. A year later, when the devastating effects of Katrina bore down on the old, inadequate levee system in New Orleans, Hanak, Lund, and their colleagues saw an indication of the future of the Delta. However, they saw an earthquake, not a hurricane.

As Lund, Hanak, and other scientists who study the Delta watched Hurricane Katrina bear down on New Orleans, they felt compelled to prevent something analogous from happening in their backyard. Scientists had long warned of the fate that awaited New Orleans if its infrastructure was not improved to prepare the city for a major hurricane. But political apathy, and perhaps human nature itself, prevented these scientists from making much headway.

Approximately 2 million people in the Central Valley count on levees for flood protection. And the capital city, Sacramento, which is among the fastest-growing cities in the United States, is the major metropolitan area at the highest risk of flooding. The problem is that Sacramento's infrastructure is inadequate; it doesn't meet even the minimal federal standards. Conservative estimates of potential flood damage to the Sacramento area alone exceed $25 billion. Sacramento is, of course, just one city among many in the Central Valley. Actually, Sacramento is well upstream of the Delta, and its land is above sea level, so failure in the Delta is unlikely to flood major population centers. Nevertheless, if the Delta goes, it can take a lot down with it. In addition to flooding large areas of lower-value agricultural land, it would cripple the delivery of water to the San Francisco Bay Area, Southern California, and the San Joaquin Valley.

So the Delta is now a big fishbowl—a fishbowl not very good for the native fish. The Delta is the habitat of more than fifty species of fish, including 75 percent of the state's commercial salmon catch. Today, the Delta supports what scientists kindly refer to as a *highly modified ecosystem*. Hanak explains, "We've engineered this system to the point where it's a lot more vulnerable. The invasive species like the artificial things we've created. But we're legally bound to protect the native species that aren't adapted to the new Delta we've created."

The Delta today is, in fact, a shadow of its former self. It resembles the Delta of the past only in that some of the original species, such as the delta smelt and chinook salmon, are still present. Invasive species, both plants and fish, now dominate the Delta's riprapped channels and islands; native species, including the delta smelt, the longfin smelt, and salmon, to put it mildly, are struggling.

The recent sharp decrease in the population of several prominent Delta fish species was a red flag for conservationists, many of whom believe that the entire Delta ecosystem is on the verge of collapse. Two species have already gone extinct in the Delta: the Sacramento perch, which needs to be reintroduced; and the thicktail chub, which has been globally extinct since 1957. Six Delta fish species are heading toward extinction: the southern green sturgeon, the longfin smelt, the delta smelt, the winter run chinook, the spring run chinook, and the Central Valley steelhead. Two species are in decline: the splittail and the late fall chinook.

With regard to global warming, every place has its own counterpart of the canary in the coal mine. In the Central Valley, the canary is most likely the tiny delta smelt, a translucent fish about the size of a human finger. In the fall of 2004, routine fish surveys registered sharp declines in the delta smelt, and it was listed as *threatened* under the Endangered Species Act. The number of delta smelt found in 2008 was the lowest in forty-two years of surveys. Federal scientists say the delta smelt is on the brink of extinction; some biologists conclude that it may be gone by 2010.

When scientists say the Delta's native ecosystem is collapsing, they mean it. A major factor contributing to the collapse is the pumps. These pumps are extremely powerful and can kill fish that get stuck inside them. In addition, the reduction of total outflow to the ocean and the disruption of natural flow patterns in what is now a complex network of channels in the Delta aren't helping matters. The situation has gotten so bad that in December 2007, U.S. District Judge Oliver Wanger imposed pumping restrictions to protect the delta smelt. The courts imposed 30 percent reductions on the amount of water that can be exported from the Delta by state and federal water projects. This means that even if water is available, it may not be delivered. On March 5, 2009, the state's Fish and Game Commission unanimously voted to list the longfin smelt, a relative of the delta smelt, as a threatened species under the California Endangered Species Act. The commission also voted to classify the delta smelt as an endangered species. (It had been listed as threatened since 1993.) The ruling to protect the environment has disrupted the water supply of much of the state.

The scientists concerned with the Delta hoped to do better than those who had been concerned with Katrina, by giving people a vision of what lay ahead. The title of their first report was *Envisioning Futures for the Sacramento–San Joaquin Delta.*[1]

"Back in the summer of 2005, I got in touch with these guys and we all agreed that someone needed to start looking at the Delta. It was a problem that required a multidisciplinary approach. There are so many moving parts," Hanak explains. "It was clear the Delta required more sophisticated long-term planning. And after years of neglect, the Delta ecosystem was showing signs that it had become unhinged. The crash of some Delta fish species between 2004 and 2005 helped focus attention."

Hanak says it was important to combine an economic perspective with an engineering perspective. "The way we try to look at it is that there are two primary objectives: water for humans and a healthy ecosystem for fish populations. If you do those two things . . . you might make the Delta more sustainable," Hanak says. "We wanted to give people a look at possible futures. We thought that it would help people to think long-term. We can't just tinker with the status quo."

————

Ask any scientists who study the Delta and they will tell you that it is likely to change significantly and abruptly within the next generation. A sudden catastrophic change would be a very hard landing indeed for those depending on the Delta. When a system pinned for more than a century swings loose, it's going to take a lot down with it. The scientists feel that their job is to help people see what a catastrophic failure would look like and then provide options to prevent the worst-case scenario.

The scientists might not be worried about a hurricane in the Delta, but they've got plenty of other scenarios that would result in a catastrophic failure. An earthquake and a flood are the two most likely scenarios, and both could take the levees out fairly quickly. According to recent calculations, the odds are roughly two in three that during the next fifty years either a large flood or a seismic event will affect the Delta. And the scientists say that these odds are a conservative estimate, for two obvious reasons. First, strain continues to accumulate on faults in the Bay Area, increasing the risk of seismic activity with each passing year. Second, the 100-year flood isn't what it used to be, thanks to global warming. The estimate of what a 100-year flood event in the Delta looks like is based on outdated hydrology data that don't adequately address the impact of climate change on the Delta. Scientists have already established that in recent years, climate change is causing much higher inflows from rivers feeding into the Delta.

When I asked Lund if it was fair to say that a catastrophic levee failure of some kind was guaranteed by the end of the century if nothing was done, he answered as any good engineer would: "Well, you can't give it a 100 percent probability. But I might put it at 99 percent."

Aside from the cause, the scientists see a lot of the same issues in the Delta that their counterparts had warned about with regard to Katrina. Lund and Hanak envision a levee failure that would directly threaten water supplies and affect thousands of roads, bridges, homes, and businesses at the same time.

An earthquake is the quickest way to take down the Delta, and even though earthquakes are geologic—not climate-based—events, the changing of the landscape through global warming will determine its resilience. It's what scientists call an unavoidable threat, and it's been on their radar for more than thirty years. There are at least five major faults close to the Delta, and these are capable of producing significant ground accelerations. The soils are poor enough and the levees weak enough that risk of failure due to liquefaction and settling is high. You can do a lot of seismic risk studies in thirty years, and all of them indicate a very high potential for major earthquakes in the region sometime in the near future. If there is a major earthquake—similar to the 1906 event in San

Francisco, which measured 8.25 on the Richter scale—many Delta island levees will fail simultaneously. And with each inch of rise in the sea level, the cost of such a failure increases significantly.

Even in a scenario involving a moderate-magnitude (6.0) earthquake, the seismic risk studies show a potential for multiple levee failures. The highest risk is in the western Delta, which is very close to several significant seismic sources and is already characterized by deep subsidence and poor foundations. There is a medium to high risk of catastrophic levee failures for almost all the central Delta as well. Scientists working for the Department of Water Resources (DWR) recently modeled the consequences of a catastrophic levee failure caused by a large earthquake. In one of the scenarios, the earthquake took out thirty levees, flooded sixteen islands, and cut off water exports for several months.[2]

But what the scientists fear most is something called the "Big Gulp." The name itself sums up the scenario. If the levees break, salt water from San Francisco Bay will come rushing in, proving that nature abhors a vacuum. Lund does a quick back-of-the-envelope calculation: "It would take as little as twelve hours for the salt water to begin intruding into the Delta."

In another earthquake scenario, scientists simulated a magnitude-6.5 earthquake that takes down twenty islands. This scenario shows a big tongue of salt water creeping in from the bay. Within thirty days, the Delta would be transformed into a saline estuary. The extra salt in the water would wreak havoc on the millions of people and the millions of acres of farmland that depend on Delta waters. Scientists estimate that a catastrophic failure of key levees would cost, in total, somewhere between $8 billion and $15 billion.

Even without catastrophic failure, rising sea levels will bring more salt into the Delta and significantly raise the cost of water treatment; raise the public health risks to the millions of Californians who rely on the Delta as a source of drinking water; and reduce the productivity of farms, which would be irrigated with increasingly salty water. An intrusion of salinity can be delayed for a time by releasing more freshwater into the Delta, as Lund explained, but it cannot be delayed indefinitely.

With regard to an earthquake, there is also the question of when. "If the earthquake happened during the summer when things are dry and water levels are low, it would get salty a whole lot more quickly," says Hanak. "All these islands are bowls and all of them would be filled with saltwater. If the earthquake happened during winter there would be more freshwater in the Delta to start with, so the Big Gulp would be a lot smaller." Timing, as they say, is everything.

Unfortunately, global warming is also a matter of timing. Global warming is insidious in that it discreetly fiddles with the timing of the Sierra Nevada snowpack. This snowpack is the true basis of California's water system. It's the state's largest surface reservoir; even though the water is in solid rather than liquid form for several months of the year. Snowmelt currently provides an annual average of 15 million acre-feet of water, slowly released between April and July each year. An acre-foot, as the term implies, is the amount of water needed to cover 1 acre 1 foot deep. More significantly, 1 acre-foot is 325,851 gallons, or about enough water to supply one to two California families for one year or to irrigate from one-eighth to one-third of California farmland, depending on the crop.

Mountain snowpack is like money in the bank. Much of California's water infrastructure was designed to capture the slow spring runoff and deliver it during the drier summer and fall months. However, warmer temperatures would cause the snow to melt earlier, thereby reducing summer supplies. Using a combination of historical data and climate models, the DWR projects that the Sierra snowpack will experience a 25 to 40 percent reduction from its historic average by 2050, thanks to global warming.

Over the past 150 years, monitored mountain glaciers have been shrinking. And with the earlier melting disrupting the timing, there are now more floods during the winter and worse droughts during the summer. A flood of salt water can of course do enormous damage, but a flood of freshwater isn't much better. Over the last fifty years, there has been a shift toward less snow and more rain in the Sierra Nevada. Climate change is also expected to result in warmer storms that bring less snowfall at lower elevations; this is more bad news for the snowpack and means, to use the metaphor above, less money in the bank for the state of California. These shifts have increased winter inflows to the Delta. Climate models indicate that this trend will continue, with even larger and more frequent floods in the future.

Whereas you might expect the melting to result in a short-term increase in the amount of water available in the "bank," the disrupted cycle is actually more likely to cause excess runoff, bringing flooding and an overflow of reservoirs not equipped to contain such large inflows of water. Steve Schneider, a climatologist at Stanford University, puts it this way: "Water managers have this horrible choice. You can leave the reservoirs full and hope you don't get a heat wave in February or March that melts snowpack early, floods productive farmland, and inundates surrounding communities."

This is exactly the kind of scenario that scientists expect. A week of unusually high temperatures leads to an exceptionally early snowmelt, which causes a big pulse of meltwater at a time when reservoirs are already full. If that week of heat is combined with a storm that brings heavy rainfall, significant flooding is likely.

Schneider continues with his explanation: "The other option is to play it safe and release water from the reservoirs, bringing the level down. Then you're praying for rain because you've got nothing to fight fires." Schneider adds, "Either way you're praying a lot." This, Schneider says, is why Californians have become deeply involved in climate legislation. "They don't want to deal with the increased risks of droughts and floods," he says, "Ideally, they'd be happier with a whole lot less climate change."

But California has already seen its share of climate change. During the last fifty years, winter and spring temperatures have been warmer, snowpack has been melting one to four weeks earlier, and flowers are blooming one to two weeks earlier. One of the fundamental problems with life on Earth is that natural resources, such as clean water for cities and for growing food, is not evenly distributed in space and time. As a result, we've tried to engineer the system to be more evenly distributed. Scientists even out the distribution by looking at past variability and building infrastructure to smooth away the bumps.

But global warming is working against us. So far, California is still betting that a system built using historical hydrological data can protect the future water supply and provide sufficient flood protection. But in the meantime, climate change has already redefined the hydrology of the Delta, making it quite clear that our past experience is not enough to serve as a guide. To think about it any differently would be dreaming.

A rising sea level makes the situation worse. If San Francisco Bay serves as the source of the Big Gulp, then a rise in sea level is the Little Gulp. During the past century, the sea level along California's coast has risen about 7 inches. It is projected to rise an additional 12 to 55 inches—or possibly even more, if large ice sheets melt—by the end of the century. Without immense investments to raise the Delta's levees, this rise in sea level will cause many levees to fail, pushing seawater into the Delta. A rising sea level can also increase the rate of saltwater intrusion into coastal aquifers; such intrusions would contaminate freshwater supplies. Even if the levees could be sustained, a higher sea level will increase the salinity of Delta waters. And the higher tide that sweeps in with a rising sea level would pose additional problems.

A higher sea level will make pumping water through the Delta increasingly unattractive and eventually infeasible. Even if the existing levee network could be maintained through unprecedented investments, the worsening water quality resulting from the rise in sea level would steadily reduce the economic value of water exports from within the Delta. The current costs of salinity in the Delta are already significant for agriculture and urban drinking water treatment in the southern Central Valley. More saline exports from the Delta will reduce the viability of agriculture in this region and increase costs of and health risks from drinking water from the Delta. With a continued rise in sea level, the volume of required outflows would continue to increase. Climate models suggest that by mid-century, the increased salinity of Delta waters will impose water treatment costs of about $300 million to $1 billion per year, every year.

Higher salinity will impose a direct water supply cost by requiring higher outflows to keep seawater away from the pumps. Scientists estimate that with a 1-foot rise in sea level, an annual average of at least 475,000 acre-feet of additional Delta outflow would have been required to maintain salinity conditions from 1981 to 2000 at the western edge of the Delta. That works out to about 10 percent of annual export volumes during the period. With an additional 3-foot rise in sea level late in this century, pumping through the Delta may no longer be practical. Steve Schneider says, "And that's why the Delta is the single most vulnerable place to sea level rise in the entire state. We've completely transformed the landscape! And we've transformed it in such a way that it has no resiliency. No one was thinking *sea level rise* when they designed this stuff. It used to be a giant marsh! Now we use it to grow alfalfa for animal feed."

But California has to play the cards it was dealt. Left to flow naturally through the state's rivers, most of the precipitation that falls in California would flow out to the Pacific Ocean either directly through the rivers of the north coast or through the San Francisco Bay via the Sacramento and San Joaquin rivers. This would leave the southern part of the state—which contains about two-thirds of the state's population—with little of California's available freshwater supplies.

Lund agrees that the Delta is at a tipping point. In 2008, the scientists I mentioned earlier teamed up again, and Lund was the lead author of a report titled *Comparing Futures for the Sacramento–San Joaquin Delta*.[3] The report looked at various options for preventing a catastrophic collapse of the levees. It was meant to offer realistic solutions. It was an opportunity to prevent what might otherwise be inevitable.

"It's a game of chicken," Lund says; "and I just hope they can stop playing chicken before the earthquake happens." Hanak adds, "The current situation in the Delta is bad for the economy and brutal for the fish. Yet folks in the Delta don't want to see any change. Our evidence suggests that is not possible. They need to be thinking proactively. And they're going to need help financially."

If the scientists' first report was an attempt to envision various future scenarios for the Delta, then their second report presented four possible ways to move the water around. Lund says, "If you want to have a significant amount of long-term water exports from the Delta and you want fish, you need a strategy. You can always screw that strategy up. You can engineer a system with the best intentions and make it worse. But if you do things right, the optimal strategy should at least be better than all the other strategies. Based on our analysis it looked like a peripheral canal was the optimal solution if you want to continue with significant exports." Hanak says, "It doesn't mean everyone agrees with us."

The proposed *peripheral canal* (a canal that goes around the periphery of the Delta) is fairly straightforward, and it isn't new. The first such proposal was defeated in a state referendum in 1982; Northern Californians turned out in force to vote against it. As a result, it remains controversial today. The scientists are now offering a proposal that improves on the original concept but conservation groups are still worried that if a peripheral canal is built, it will allow for even more unsustainable levels of water exports to the south. They might be right. As Lund says, you can screw up anything.

But if we don't screw it up, the canal would protect water exports in the event of a levee failure and would also help reduce the extent to which water flow is altered by the current configuration of the pumps. Pumping water directly out of the Delta clearly hurts the environment. Water diversion pumps in the southern Delta are so powerful that they actually make the Delta's maze-like channels flow backward. Hanak explains it this way: "You're not sucking water through the Delta to the pumps. The peripheral canal decouples the management of water for humans with the water for the Delta ecosystem. You bring the clean water directly to the pumps and you don't have to keep the Delta fresh. It'll get saltier in the fall. We pump all year round right now. You're also not sucking the fish down. We've been operating this system and pumping so much the rivers go backward. You're not as vulnerable to a catastrophic levee failure. You just have to repair the canal, not the entire levee system."

The scientists see the canal as an essential way to separate the state's water demand from a Delta environment under grave stress. Sixteen Delta fish species are being pushed toward extinction, in part, by this demand.

"If you're just worried about fish," Hanak explains, "you want absolutely no water exports. But the engineering question is: how much water can you take out of the system and not hurt the fish? If you're optimizing over both human needs and ecosystem needs, you'll see reductions in uses for things like low-margin agriculture. It will force us to cut back on lands that are least productive. It's hard to do. And that's why this is one of those things that has been off the table for a long time. The Bay Area actually depends more on the Delta than Southern California does. Most of the growth in the Bay Area has relied on water from the Delta. If Bay Area folks understood the implications, then I don't think they would object."

The peripheral canal would deliver water from the Sacramento River along the Delta's eastern edge and then down to the pumps in the south, in effect circumventing the Delta itself. That's why it is called peripheral. It requires building an earthen canal wider than two football fields and more than 40 miles long. The canal would give the Bay Area and Southern California a direct line to tap into the Sacramento River. The channel would divert some of the river's flow around the fragile Delta and on to existing pumps in the southern part of the Delta. From there, the river would continue to serve Los Angeles, San Diego, farms in the San Joaquin Valley, and much of the Bay Area.

Originally, the peripheral canal was seen only as a water conveyance. Now, many scientists also view it as a restoration tool and a hedge against disaster. But residents along the eastern edge of the Delta are concerned about the effects of a 40-mile canal with a footprint 1,000 feet wide. For this reason, scientists say that any consideration of a canal must first begin with a commitment to water conservation and efficiency efforts on a scale not yet attempted in California. Hanak says the price of the peripheral canal would be somewhere between $5 billion and $10 billion.

"That's not such a big problem," she says. "Water users can cover that. The current system is so unreliable and risky. The bigger question is who pays for improvements in fish habitat and for the mitigation for farmers who rely upon the Delta to make their living."

Lund adds, "It's about adaptive management. We don't really have any other choice when you begin to factor in the impacts of climate change, like sea level rise, and how these new forces will be pushing on the system. If we go ahead and build the peripheral canal, we've essentially constructed a whole new Delta. We're not going to know exactly how to operate it."

Adaptive management is an admission that pinning a situation down will never work, but it requires a willingness to make changes and to be flexible. "It's real water and it's real money to someone," says Lund. "It's very hard to do adaptive management because every experiment is real water and real money. But we need to do this."

That said, Lund still thinks the more likely possibility is that, as in the case of Hurricane Katrina, the decision regarding what to do will be made for us, something Lund calls "failing into a solution."

According to Lund, "The physics is not going to wait for the policy to be put in place. If you have the big earthquake, the event forces you into making the quick policy decision. And that quick

policy decision could be a complete disaster. That's what I call failing into a solution." When I asked Lund if he thought it was likely
that a good policy decision could be made in advance of the worst-
case scenario, he wasn't necessarily optimistic. "It's incredibly difficult to reach a consensus. It asks too many people to give up too
much."

But if an adaptive style of management were to go into effect,
it could also deal with other big problems—the kind of problems
that come with living in a place subject to extreme weather and
extreme climate. For example, El Niño—the periodic warming of
the Pacific Ocean—routinely inflicts millions of dollars' worth of
damage on California. Losses caused by storms and floods during
the 1997–1998 event alone were over $1 billion. Adaptive manage-
ment would help reduce the state's vulnerability to El Niño events.
Schneider says climate change is another problem that could be addressed. "The two big problems you need to keep in mind when you
think about climate change in California," he says, "are wildfires
and pollution."

California has the worst air quality in the United States. More
than 90 percent of the population is in areas that violate the state's
air quality standard for either ground-level ozone or airborne particulate matter. These two types of pollutants can cause or exacerbate a
wide range of health problems, including asthma and heart attacks.
They have also been shown to decrease lung function in children.
Combined, ozone and particulate matter contribute to 8,800 deaths
and $71 billion in health care costs every year. The connection with
global warming is nothing more than simple chemistry. Higher
temperatures increase the formation of ground-level ozone and particulate matter. Ambient ozone also reduces crop yields and harms
the ecosystem. If global background ozone levels increase as projected in several future climate scenarios, it may become impossible
to meet local air quality standards. Air pollution events will become
more frequent, longer-lasting, and more intense as temperature increases. For example, scientists at the California Climate Change

Center forecast that the scenario based on moderate temperature increase almost doubles the number of days with weather conducive to ozone formation in the San Joaquin Valley, relative to today's conditions.

And then there are wildfires. Warmer temperatures, drier conditions, and increased winds could mean hotter wildfires that are harder to control. Aside from their destructive potential, wildfires also increase levels of fine particulate matter. Therefore, air quality could be further compromised by increases in wildfires, which emit fine particulate matter that can travel long distances, depending on wind conditions. The most recent analysis suggests that if greenhouse gas emissions are not significantly reduced, large wildfires could become up to 55 percent more frequent toward the end of the century. The California Regional Assessment notes an increase in the number and extent of areas burned by wildfires in recent years; and modeling the results under changing climate conditions suggests that fires may be hotter, move faster, and be more difficult to contain in the future.

———

When I ask Hanak what the Central Valley might look like by 2050 if we do nothing, she says, "The Delta water system will not be viable. You'll see a lot of open water, and the folks who depend on that water won't have access to it. We're likely to lose 1 million acres of San Joaquin farmland. It wouldn't be a devastation to the state economy, but it would certainly hurt that region, one of the country's most productive farming areas." In 2009 the California Climate Change Center released a report that builds on this picture. According to the report, water shortages across California are more likely in the future as snowmelt decreases, the climate warms, and the population grows. California's population is expected to grow from 35 million today to 55 million by 2050.

Climate models suggest that by the end of the century, late spring stream flow could decrease by one-third. Agricultural areas could be hard hit, with California's farmers losing as much as 25 percent of the water supply they need. Climate models indicate that by 2050 the average annual temperature in California will rise between 1°F and 2.3°F, depending on the level of global greenhouse gas emissions. And by the end of the century, if emissions proceed at a medium to high rate, temperatures in California are expected to rise somewhere between 4.7°F and 10.5°F. In a more optimistic scenario, with lower greenhouse gas emissions, temperatures still go up between 3°F and 5.6°F by 2100. If temperatures rise to the higher warming range, there could be as many as 100 more days per year with temperatures above 90°F in Los Angeles and above 95°F in Sacramento by 2100.[4]

In 2010, the peripheral canal will be under the spotlight as the proposal reaches the governor and the legislature. There will be lots to wrangle over, including whether and how to build it, as well as how the state will buy up land for 100,000 acres of environmental restoration in the Delta. A committee of government agencies is working on a related habitat conservation plan for the Delta, also due in 2010, which is expected to include a canal. The state Department of Water Resources is drafting an environmental impact report on canal options, also due for completion in 2010. Lund says, "You never know what will come out of the sausage machine that is politics." But what seems clear is that a classic Hollywood ending is unlikely.

Hanak and Lund will tell you that in order for the Delta to survive, it will need a substantial overhaul. They can offer what they see as an optimal solution, but they can't offer a perfect solution. The optimal solution means that some fish will still die and some farmers will lose their livelihood. It is not—to repeat—a Hollywood ending. The problem with engineering is that it provides a methodology for calculating your losses ahead of time. That means you have to take responsibility and accept those losses. Lund is not sure rational solutions are the inevitable choice, because people would rather simply hope for a happy ending. "The most likely outcome is we will not be able to decide in time. And our reports will be used as a hind-cast instead of a forecast. Our studies will serve as a nice example of what could have been done. And we will forever be known as the Katrina scientists of the Sacramento–San Joaquin Delta."

## The Central Valley, California: The Forty-Year Forecast—Drought, Water Resources, and Agriculture Problems

| SACRAMENTO, CALIFORNIA | TODAY | | 2050 | | 2090 | |
|---|---|---|---|---|---|---|
| Emissions Scenario | JAN. | JULY | JAN. | JULY | JAN. | JULY |
| Higher | 45 | 76.1 | 47.5 | 80.2 | 50.5 | 84.3 |
| Lower | | | 47 | 79.3 | 48.3 | 80.8 |

## Forecast
## April 2017

By 2017, the future began to reveal itself to the residents of California. Temperatures continued creeping up, winters were becoming increasingly tepid, and spring came earlier and earlier. Summer seemed to last forever, whereas the Sierra snowpack was gone by June. None of this came as much of a surprise. It had always been fairly easy for the climate models to forecast temperature. Mostly, we were just relieved that it wasn't worse. At least the seasons were still pretty much the seasons, even if they were all tracking warmer than they should be. Those all-important winter storms that blew in from the North Pacific Ocean bringing rain and the occasional snowfall, part of the classic so-called Mediterranean seasonal precipitation pattern, were still around. We figured that as long as the timing of the rains remained about the same, California could focus on dealing with the heat.

It was in 2017 that a storm gathering in the Pacific would redefine how we looked at winter. It would also redefine how we looked
at the ocean-atmosphere phenomenon called El Niño, a periodic
warming of the equatorial Pacific Ocean. El Niño was now further
juiced by all the extra heat in the atmosphere, and the winter storms
it brought were expected to knock some of California's rainfall records out of the ballpark.

For the most part, El Niño events typically lasted about one
year. But coral records from Palmyra Island in the tropical Pacific
Ocean suggested that there were periods in the past when both the
amplitude and the frequency of El Niño events changed abruptly.[5]
Whatever the cause, the present El Niño seemed to be different. As
the climate models struggled to describe how global warming might
influence El Niño events in the future,[6] we watched as it changed
right before our eyes.

The warming of sea surface temperatures in the equatorial Pacific was evident by early April 2017. And in early June, the NOAA
Climate Prediction Center issued an El Niño Advisory stating that
ocean temperatures in the central and eastern Pacific Ocean were
the warmest since August 1997. Until now, the 1997–1998 El Niño
had been one of the strongest in recorded history. It was lucky for
California that we had learned from past mistakes. Following the
large El Niño of 1982–1983, an event that went largely undetected
until it kicked California in the teeth, federal money was used to
make improvements to the observational network that fed into the
climate models. With the installation of the TAO/TRITON array
of moored buoys, scientists finally had more comprehensive data
regarding the condition of the tropical Pacific Ocean. The data collected by means of this improved observing system supplied the climate models with very high-resolution snapshots of ongoing ocean
conditions. The models, in turn, predicted that a strong El Niño
was now growing in the Pacific and that it would continue to grow
through early 2018 and perhaps even longer—with emphasis on the
word *longer*.

By late spring 2017, scientists at NOAA issued a climate forecast for the approaching winter months, highlighting El Niño's probable impact on temperature and precipitation patterns in the United States. Mother Nature's forecasts, often more subtle, were not far behind. Luckily, there were plenty of old-timers, knowledgeable about wild creatures and the local environment, who were still capable of reading these more subtle messages with great skill. Phone calls came pouring in to the Marine Mammal Center in Sausalito about sightings of stranded sea lions. The sea lions, unable to find enough anchovies or herring, were malnourished. It was likely that the warming ocean temperatures associated with the growing El Niño were driving the fish away.

A lack of upwelling of cold water along California's coast was the key to many of the strange sightings. Wind-driven upwelling of cold, nutrient-rich deep water acts as a feeding trough. But El Niño turns down those winds, and that downturn reduces the upwelling. Reduced upwelling prevents nutrient input to surface waters, lowers the amount of plankton—tiny marine plants and animals—and messes with nearly everything that relies on the ocean for sustenance. Biologists studying salmon in the waters off San Francisco reported finding razor-thin sardines and anchovies, underweight for their size and probably undernourished. A brown pelican was spotted in a suburb of Phoenix, probably propelled by El Niño winds from the California coast all the way to Arizona. An official from the Maricopa Audubon Society explained that many starving birds had just given up and let the wind carry them.

According to NOAA's El Niño Watch, tropical El Niño conditions would continue to gain strength through January along the U.S. west coast, with sea surface temperature (SST) departures similar to those observed in December. The SSTs were 5°F to 6°F above normal off the northern California coast, more than 6°F above normal off the coast of San Francisco and Southern California, and 4°F to 5°F above normal off the coast of the Pacific Northwest.

By February the worst of the winter brought heavy rains, flooding, and mudslides in northern California, and there were flood warnings for the Russian and Napa rivers. Interstates were shut. Heavy surf, rising to 50 feet, eroded beaches. As much as 14 inches of rain fell in Santa Barbara. Hurricane-force winds of up to 95 miles per hour caused devastation. Residents of Sacramento stockpiled sandbags and cleaned out channels. And the 2 million people of the Central Valley who were counting on the levees for flood protection heaved a sigh of relief. The levees, for the most part, had survived the storm intact. But the overall frailty of the levee system was in plain sight, and the storm revived stalled discussions about the need for a peripheral canal. It was clear that something had to be done, and it was hoped that this El Niño might serve as the impetus.

After the first El Niño dissipated, climate models forecast one more round. In other words, the El Niño of 2017–2018 was technically the first of a pair. Surfers in Newport Beach had a field day. But while they were riding the best winter waves of the century, California itself was riding headlong into an economic meltdown. This time it was caused by a flood of water as opposed to a flood of bad mortgages. The state, still suffering from the bursting of the real estate bubble in 2009, was already strapped for cash and ill-equipped to make any significant infrastructure upgrades. There was talk of trying to float a $10 billion bond for the peripheral canal. But in the end, no one had the cash to support it.

There was one small gift amid all the wreckage. The El Niño rains generated a blooming of spring flowers across Death Valley. The winter storms that brought mudslides and death to Southern California dropped more than 10 inches of rain on this thirsty desert—five times more than usual—persuading the hearty wildflower seeds to blossom year after year. Death Valley hadn't seen such an array of flowers since the spring of 2005, almost fifteen years earlier. In March and April 2017, tourists came from around the world to see the landscape repainted in the rich blue of the desert lupine, the royal purple of the chia flower growing in clumps along the side of the road, the magnificent golden yellow of the California poppy and the desert dandelion, highlighted by the soft white petals of the tufted evening primrose scattered across the hillsides.

## August 2027

By 2027, the dream that was California had begun to unravel. Temperatures across the Golden State, which had been rising since the 1950s, started to spike. August 2027 saw new record high temperatures being set almost every day. As one local meteorologist put it, "The good news is that the number of official heat waves in Sacramento this year is finally down, the bad news is that the heat wave that started in July is still going strong." Sacramento, along with much of the U.S. Southwest, was trapped in one long heat wave.

The intensity of this heat wave was crushing, and the duration of abnormally high maximum and minimum temperatures was reaching into every sector of California's economy. An all-time record for statewide energy consumption was set, and hundreds of heat-related deaths were reported. Meteorologists watched in awe as the semi-permanent four-corners high-pressure system expanded to the west and north. Associated with this expanded high-pressure system was an unseasonably warm air mass settling over the region through the duration of the event, with an influx of monsoonal moisture from the Gulf of California and the Baja region. Typical mid-level winds from the southwest shifted to the southeast. This allowed for maximum transport of moisture over southern and central California, and served to lift minimum overnight temperatures higher. It also brought excessively humid conditions during the daytime hours—a situation not typically associated with California. In other words, California felt like Louisiana.

The heat wave proved devastating to the state's $3.2 billion wine-making industry. California, still the nation's largest wine producer and the fourth-largest wine producer worldwide, with its high-quality wines produced throughout the Napa and Sonoma valleys and along the northern and central coasts, had, in a sense, come under siege. High temperatures during the growing season caused grapes to ripen prematurely and reduced their quality. Specifically, an increase in the overall number of temperature extremes above 86°F during the growing season had shut down photosynthesis. Cooler spots, such as Mendocino and Monterey counties, were still hanging in. But the grape-growing regions of the Central Valley, with its already marginal conditions, were hit hard. California's losses due to heat were estimated at 30 percent and close to $1 billion.

It was a summer that seemed like science fiction and apparently wasn't set on the planet Earth. Mudslides became a big problem in areas deforested by wildfires, and these slides were almost impossible to escape. Airports were virtually forced to shut down as plane after plane was grounded, week after week, by decreased visibility from the wildfires. Air quality became unbearable. The entire state violated air quality standards for ground-level ozone (smog) and small particles. Air pollution in Los Angeles and the San Joaquin Valley was especially bad. The cost of managing an unruly climate had become crippling.

The models were right: over the past two decades, California's average temperature had risen somewhat more than 2°F. The temperature increase was further proof that a high-emissions scenario was happening. This meant that by mid-century, the serious changes would to start to kick in. And if nothing was done to ramp down emissions, by the end of the century, climate models projected that statewide average temperatures would rise more than 10°F. We couldn't even manage 2°F.

## October 2040

By 2040, the seasons had become almost unrecognizable. The heat hung on as summer extended into a long, mind-numbingly brutal test of patience. Rainfall had become erratic, and a cool patch of tropical Pacific Ocean temperatures suggested that we were locked in something like a prolonged but sketchy La Niña event. For the U.S. Southwest, La Niña is synonymous with drought. Water was on everyone's mind.

Ironically, the original pact that governs water supply in the West was negotiated in 1922, during one of the wettest periods of the past 1,200 years. Paleoclimatologists, analyzing tree ring records across the Southwest, said it was only a matter of time before the western United States reverted to its old ways and the Colorado River became drier again. We knew that global warming would push the system back into drought harder and faster.

Climate change was indeed a game changer, especially for utilities and water managers who were trying to make sure everyone had a fair share. The pact of 1922 that allocated Colorado River water to California, Nevada, Arizona, Utah, Colorado, New Mexico, and Wyoming had come about in a very different world and did not fit readily into this new one. There was still plenty of argument about whether the new reality was in fact new or just an old reality come back to haunt us. But in any case, the reality was that it wasn't raining.

As the seasons became unfamiliar, so did the landscape. Its color and texture had begun to shift before our eyes. To start, high-elevation forests in California were receding. They just couldn't take the heat. Almost half were gone, replaced by grasslands. The desert, once lush with wildflowers, was being slowly overtaken by alien species such as red brome and buffle grasses—plant species native to Africa and the Mediterranean and able to do well in high temperatures. Not only did these noxious weeds outcompete some native species in the Sonoran Desert; they also fueled hot, cactus-killing fires. The magnificent saguaro cactus (*Carnegiea gigantea*)—the state flower of Arizona—and the Joshua tree (*Yucca brevifolia*) had become hard to find.[7] In fact, the majority of California's native species, numbering well over 3,000, were fading. It seemed as if everything was heading for the hills—but the hills were on fire.

One of the first lessons regarding climate change is that the conditions of the past can't predict the future. The Hoover Dam, completed in 1936 and once a proud symbol of American engineering power and vision, became a sorry reminder of our collective inability to change with the times, let alone stay one step ahead. Straddling the border between Nevada and Arizona, Hoover Dam was built to tame the flow of the magnificently wild Colorado River, which stretched over 1,450 miles from Colorado's Rocky Mountains to the Gulf of California. The Colorado River had supported the growth of cities such as Las Vegas and Phoenix, providing about 30 million people with water for drinking and irrigation. By the year 2040, Hoover Dam was empty, and a sort of bathtub ring—a thick white band of mineral deposits—marked the walls of Black Canyon. That ring showed where the waterline used to be—before the rain stopped but while the people kept pouring in.

Before the last drops were released a few months ago—to Las Vegas, Los Angeles, San Diego, Phoenix, Mexico, and some other places—our strategy had become hope and prayer. We hoped and prayed that life would just go back to the way it used to be, when there was still rain and we didn't have to think about water all the time. Casinos like the Bellagio and Mirage, with their beautiful fountains, lagoons, and waterfalls, reinforced our hope and made it easy for us to think that everything would be fine. My favorite was Planet Hollywood, where it rained every hour on the hour, day after day. The daily rain show became one of the most popular attractions—it almost seemed to prove that Sin City could not possibly be in the midst of a water crisis.

But Lake Mead and Lake Powell were always the biggest poker tables in Nevada. Back in 2009, water resource managers at the

Southern Nevada Water Authority were asking all the right questions about the cards they were holding. Was the drought one of the traditional droughts that the Colorado River had experienced in the past, or was it something very different? In just two weeks in April 2009, managers watched as Lake Powell lost the equivalent of 14 feet of snowpack. And by the summer of 2009, the reservoir's water level fell to its lowest point since 1965—back when Lake Powell was new and officials first diverted water from the Colorado River into it. By the end of August 2009, Lake Mead's elevation teetered at just 1,092 feet above sea level. Water resource experts knew they were betting on the future of the West. A mere 17 feet stood between hope and despair. By law, once the level dropped below 1,075 feet, the Southern Nevada Water Authority was required to find alternative sources of water. The future of water in the West, the future of great bottles of cabernet and zinfandel, the future of beautiful apricots and almonds and avocados, the future of life itself, depended on that 17 feet.

The models had been predicting that climate change could reduce the runoff that feeds the Colorado River between 5 and 25 percent by the middle of the century.[8] But there was a big difference between reduced runoff and empty reservoirs. That difference was a matter of managing the double whammy of climate change and population growth. One problem was that even under the most extensive drying scenario—a drop in runoff of 20 percent—water supplies wouldn't be affected immediately, because Lake Mead and Lake Powell, which could store up to 50 million acre-feet of water, provided a significant cushion. In fact, the total storage capacity of all the reservoirs on the Colorado exceeded 60 million acre-feet, almost four times the average annual flow on the river itself. As a result, scientists calculated the risk of complete reservoir depletion as low through 2026, making it easy to postpone action. But after that, all bets were off, and the risk soared. Good management could reduce the risk, but a policy of business as usual meant that under a worst-case climate scenario, by mid-century there was one chance in two of empty reservoirs in any given year. The scientists cautioned that if the water managers took aggressive steps to reduce downstream releases during periods of drought, it would still be possible to cut their losses by as much as one-third. But they warned that if the managers did nothing, as the West continued to warm, a drier Colorado River system could have a risk as high as one chance in two of completely depleting all of its reservoir storage by 2050. At the time, it was anyone's guess when or if Lake Mead would reach that point.

Looking back, the water resource managers did the best they could, considering that the country was in the midst of a recession. The national economic downturn reduced the utility's revenues just as they came to see the risk of climate change and wanted to take steps toward new infrastructure investments. Even so, they managed to allocate $800 million and embarked on a large-scale construction project to build a new intake pipe to pull water from Lake Mead. This project, known as the *third straw*, was a way to ensure that the utility could siphon water if the lake level dipped below 1,000 feet, the point at which the two existing intakes became useless. It was an excruciatingly difficult project, but they got it done. And it bought us extra time. They also sought permission to build a controversial $3.5 billion pipeline to transport groundwater from ranches in rural eastern Nevada. But the pipeline plan drew so much ire from ranchers and environmental groups that it never made its way out of court.

## July 2050

Despite the fact that scientists had run countless simulations of an earthquake in California's Central Valley, the first twenty-four hours of the real thing were terrifying. It was a magnitude-6.5 earthquake, and it took down twenty islands in the San Joaquin Delta. As the levees collapsed, salt water from San Francisco Bay rushed in. The islands, now acting like bowls, filled up with salt water. It took a little more than thirteen hours for the salt water in San Francisco Bay to begin filling up the Delta. Unluckily, the earthquake happened during the summer, when the ground was dry and water levels were low. Just as predicted in an earlier scenario, within thirty days the Delta became a saline estuary; salt water ravaged millions of acres of farmland that had come to depend on freshwater from the Delta; and a catastrophic failure of important levees cost more than $16 billion. The president declared the Central Valley a federal disaster area. Federal and local lawmakers vowed to rebuild, and plans for a peripheral canal were finally put on the fast track.

# 8

# THE ARCTIC, PART ONE: INUIT NUNAAT, CANADA

Shari Gearheard had just returned from the 2009 Nunavut Quest dogsled race when I finally managed to catch up with her.

"I've now completed the Master Ninja course in sea ice travel and I'm 2 inches shorter for it," Gearheard said, laughing. Master Ninja is her description of an amazing six-week trip that took her 1,500 miles around Baffin Island. Her husband drove their dog team in the 400-mile race from Arctic Bay to Pond Inlet; and Gearheard, as part of the support crew, drove ahead on a snowmobile. This mix of tradition and technology—a dogsled and a snowmobile—defines life in the Arctic today. And it's safe to say that over those six weeks Gearheard saw, and felt, just about every kind of sea ice. "It gets pretty bumpy out there sometimes," she says. Like a lot of things in the Arctic, the sea ice is far more complex than it seems to be at first glance; this is why Gearheard finds it best to study the Arctic up close and in person.

If there were only two types of people in this world, summer and winter, Gearheard would definitely qualify as a winter person. "It started in childhood," she explains. "I loved winter and I loved the snow. When I was little, I used to get in trouble from my mom because one of my favorite things to do was to go by the side of our house where these big snowdrifts would form and I would just dig a big snow cave into the drift and go to sleep. She always worried that I would get trapped or something." Gearheard, now in her thirties, is a research scientist at the National Snow and Ice Data Center (NSIDC) at the University of Colorado in Boulder, where I suspect everyone else is a winter person, too. Gearheard telecommutes to NSIDC from Nunavut, in the Canadian Arctic, where she and her husband, Jake, live full time.

In Inuktitut, one of the local Inuit dialects spoken in Nunavut, Inuit means *the people* and Nunavut means *our land*. Nunavut is the largest and perhaps best-known of the four Inuit territories that make up the Canadian Arctic. Collectively, they are called the Inuit Nunaat. The other three territories are Nunavik in the northern portion of Quebec, Nunatsiavut along the coastal region of Labrador, and the Inuvialuit Settlement Region in the Northwest Territories. Gearheard's research is focused on collaborating with Inuit communities to document their knowledge of the environment and environmental change, and to link that knowledge with science. Part of her work includes understanding the impact of climate change on communities and how the Inuit are responding. You might say the dog team is part of her research, but it has also become part of her life.

"I've always loved sled dogs," she says. "When we moved here, good friends who have a dog team would take us out with them. And sometimes they would let us drive their dogs. We absolutely loved it." Gearheard and her husband thought about getting a team, but were intimidated by the amount of work involved. "Everything is made by hand," she explains: "the harnesses, the leads, the dog whip, along with learning all the commands. It's a major commitment." But one day, those same friends showed up and presented them with Siqaliq, a beautiful Inuit sled dog with an exceedingly large belly. "They said, 'Here you go; your team is going to be born in about two weeks. So you'd better get working,'" Gearheard says, laughing. And with that, she and her husband got to work.

They fixed up a dog pen and house for the puppies. Another

elder in the community gave them two more dogs; they purchased
two from a local hunter; and soon their dog team was on its way.
"Puppies grow so fast. So all of a sudden we had ten hungry dogs to
feed," she adds. And needless to say, there's no Pet Smart in Nunavut. "Our dogs eat seal meat, so we've become novice seal hunters,"
Gearheard explains. "It's our life now. All of our free time is spent
with our dogs." The dog team now has twenty stunning Inuit sled
dogs, called *gimmiit* in Inuktitut; this is the same breed used by
early Inuit to cross the Bering Strait 1,000 years ago. Actually, there
are twenty sled dogs and one seal hunting, sea ice Master Ninja.
Not bad for a kid from, as the Inuit say, *down south.*

For more than 5,000 years, the Inuit have occupied the vast
territory stretching from the shores of the Chukchi Peninsula of
Russia, east across Alaska and Canada, to the southeastern coast
of Greenland. People from down south have come and gone over
time, mostly to take things like whale, white fox, copper, or oil back
home with them. In the meantime, the Inuit have taken this cold
place, which sometimes seems barren and endlessly dark, and made
a home out of all the ice and snow.

Inuit history is preserved and passed down through an oral tradition, the telling of stories. On the basis of this tradition, the Inuit
believe that about 5,000 years ago, what they call the Sivullirmiut,
or *first people*, began to move east from Alaska.[1] (Archaeologists
use the terms *Pre-Dorset* and *Dorset* to describe these ancient Inuit
people.) In less than 1,000 years, these early people traveled across
the ice from the north coast of Alaska, across Canada, all the
way to southern Greenland. Their camps were located in places
where the hunting was good; and the bones they left behind suggest
that the Sivullirmiut hunted seals, walrus, caribou, and ducks. Depending on the season, they collected clams, mussels, fish, seaweed,
bird eggs, and berries. The Sivullirmiut also used delicate needles,
made of bird bones, to make boots and clothes from the skins of
seal, caribou, and polar bears. They made lamps from soapstone
and used these for heat and light; perhaps they even cooked meat in

soapstone pots. Unlike the Inuit of today, these early people didn't have dogsleds or large boats. Without dogsleds they were unable to cover long distances, and without large boats to hunt whales, their communities remained small.

By about A.D. 1000, the Thule people developed the large open skin boat known as the *umiaq*, as well as the harpoon that allowed them to begin hunting whales (or at least they are credited with having done so). The Thule people also used dogs and dogsleds for long-distance travel across the ice. Inuit tradition represents thousands of years of cultural developments through which the Inuit have come to master the very tough Arctic climate. Where many of us might see only vast sheets of ice, the Inuit saw endless possibilities. Lacking trees, they built igloos for winter housing out of snow, and they burned whale and seal blubber both for fuel and for lamps. They stretched sealskins over frameworks to build kayaks as well as umiaqs big enough to take out into unprotected water to hunt whales. The Inuit are the most flexible and sophisticated hunters in Arctic history and have adapted to shifts in climate with very little difficulty. This is why Gearheard and others are so interested to see how the Inuit, an ancient people who have built their lives around ice and snow, are responding to climate change.

Shifts in climate are not unfamiliar to the Inuit. It was during the Medieval Warm Period, from about A.D. 800 to 1300, that the Thule pushed east into northwestern Greenland from Canada. They were probably following the bowhead whale as the sea ice that had permanently closed off the channels between the northern Canadian islands during colder times finally began to melt in the summer. After the Medieval Warm Period came the Little Ice Age, and by the 1400s this brief return to colder conditions was well established. The Little Ice Age proved to be of little significance to the Inuit. The extra ice simply presented them with the opportunity to hunt ringed seals. In the meantime, at the Norse settlements in Greenland, settlers were still trying to grow hay and graze livestock.[2] Their inability to adapt to a changing climate ultimately proved fatal. Those early Viking colonies on Greenland have long since disappeared, suggesting that those of us from down south may have a thing or two to learn about resilience and adaptability from the Inuit.

Despite its harshness, the climate was never really a problem for the Inuit. It was the disease and alcohol brought by southern whalers and fur traders that almost took them out. Encounters between the Inuit and Europeans began in the late 1500s, when the first explorers sailed into the frigid waters of Davis Strait, Hudson Strait, and Hudson Bay. Although these first encounters were few and far between, they mark the eventual transition into what the Inuit call the *period of contact*—the period of *taking*. But the Europeans also *brought* a few things with them. Starting in the 1700s, the whalers and missionaries began to make their way north, and by 1850 they had become an almost permanent presence in the Arctic. The year-round settlement brought diseases such as smallpox and tuberculosis, which killed so many Inuit. The religion brought by the missionaries also left an impact on the Inuit. During his stay in the Frobisher Bay area in 1861–1862, the American explorer Charles Francis Hall wrote about the health of the Inuit and issued his own forecast:

> The days of the Inuit are numbered. There are very few of them left now. Fifty years may find them all passed away, without leaving one to tell that such a people ever lived.[3]

Needless to say, this forecast proved incorrect. But after a few centuries of contact, the Europeans had not only pushed the whale to the brink of extinction but also pushed the Inuit. As whales became harder to find, they were no longer profitable for commercial whalers; and for the Inuit, an important traditional source of food was endangered. Luckily, the market for whale oil and ambergris used in lamps and perfume was being replaced by a market for kerosene and synthetics. And as a result, some whaling captains and crews turned to trapping the arctic fox.

Today, through a process that began on April 1, 1999, with the establishment of Nunavut, the Canadian Arctic is officially recognized as the home of the Inuit in Canada. And since the establishment of Nunatsiavut in 2005, all four of the traditional Inuit territories have been covered by land claim agreements that establish regional autonomy within Canada. The Inuit may finally be in control of the Arctic; but as it turns out, the Arctic climate is being influenced from points farther south.

With only 152,000 people across the Arctic and limited industrial activity, there is little the Inuit can do to slow or stop global warming, because they contribute so little to total global greenhouse gas emissions.[4] But nonetheless, they are feeling its impact. The Inuit may have been able to withstand the smaller climate changes of the past, but what's happening in the present is a different story altogether.

———

Until recently, little was known about Inuit perspectives on climate change. In the mid-1990s, after speaking with elders and surveying the scientific literature, Gearheard found only a few references to Inuit knowledge of climate change. Also, she had begun to notice that although some researchers documented the impact of climate change on subsistence hunting practices, they didn't try to find out how community members *felt* about it. So Gearheard spent the next several years traveling back and forth between her home in the south and the Inuit communities of Igloolik, Baker Lake, and Clyde River in Nunavut. Through her conversations with hunters and elders—first using a translator and eventually starting to communicate on her own, speaking Inuktitut—she began to learn about Inuit knowledge and observations of weather, climate, and climate change. During one of her conversations, an elder in Igloolik, Zacharias Aqqiaruq, described the weather as *uggianaqtuq*. This Inuktitut word was later explained to Gearheard by a worker at the local research center:

> For example, I'm very close with my sister. Say I wasn't feeling myself one day and I went to go visit her. As soon as I walk in the room, or say something, she would know right away that something is wrong. She would ask me, "Is there something wrong with you?" She would say I was *uggianaqtuq*. I was not myself.
>
> —T. Iyerak, Igloolik, 2000[5]

Later, Gearheard would hear other opinions about the definition of *uggianaqtuq*. For example, it was said to be a reference to people fighting, tension, extreme heat, or something unseasonable or untimely. The root of the word may refer to a dog taking something in its mouth and shaking it. Another suggestion is that the word refers to something being eaten by lice. Though there are many suggested meanings for the word—*unexpected, unfamiliar, fighting, tension, unseasonable, untimely, being ripped apart*—interestingly, they all connect in some way to the changes Inuit have been experiencing in their environment in recent years. This was something Gearheard would hear again and again. The weather had become a stranger; it was no longer itself.

In 2004, after having worked with Inuit communities for a decade, Gearheard and her husband, Jake, made the decision to move north to Baffin Island and experience these changes and work with the Inuit firsthand. They live in Clyde River—Kangiqtugaapik in Inuktitut—a small Inuit community of about 850 people, and they are among the very few non-Inuit inhabitants who live there. In fact, they were the only non-Inuit team in the Nunavut Quest dogsled race. Clyde River is about 280 miles north of the Arctic Circle and is surrounded by some of the most dramatic fjords and cliffs in the world. The Arctic Circle is an imaginary line in latitude separating hard-core winter people from the rest of us. It is where the sun remains above the horizon at the summer solstice and remains below the horizon at the winter solstice. The Latin word *solstice* roughly translates into *sun stand still*. For Clyde River, that means the sun drops below the horizon around the middle of November and does not reappear again until the end of January. In December there is only about one hour of twilight each day. But come spring, Clyde River receives twenty-four hours of sunlight from the end of May until the end of August. It must be an amazing place in July, when the average high temperature is about 47°F and the sun never sets. But December, when darkness sets in and the average high is about -6°F, would be a different story.

Gearheard and her husband provide an interesting study in contrast. "There are some social issues in the community that are really tough," Gearheard explains. And it is these issues that her husband works on with other community members at Ilisaqsivik Society, a wellness and family resource center started by the community itself more than ten years ago. Issues such as substance abuse, domestic violence, poverty, and suicide are not uncommon in Inuit life. "Ilisaqsivik focuses on developing people's strengths and providing a lot of counseling and healing programs," Gearheard explains.

Compared with the longer timescales associated with climate change, these issues confront Inuit communities on a daily basis, and Gearheard is careful to keep her perspective. "Compared to something like, 'I can't feed my kids today,' climate change doesn't feel very immediate," says Gearheard. "Although I think people here care about climate change a lot, they definitely separate what is natural climate variability and what is not. I have often heard an elder say, 'We always had years when the sea ice was late or the sea ice broke up early, but it didn't happen eight years in a row.'"

What [I] have noticed . . . in the last five to eight years, [is that] when it should be freezing up . . . it becomes overcast, snow starts falling for a long period of time . . . that affects freeze-up . . . whenever it's overcast the temperature rises a bit, freeze-up doesn't occur as quickly.

—N. Arnatsiaq, Igloolik, 2004[6]

The climate isn't the only thing changing in the Arctic. Within a single lifetime, the Inuit have gone from living off the land to a wage economy. They are attempting to balance two very different lifestyles: one of sled dogs and subsistence hunting, the other of skidoos and soda pop. Almost all of us can relate to this clash of tradition and modern convenience.

"The food here is insanely expensive," says Gearheard. "A loaf of bread can cost $8; a small box of Tide costs $35." Much of the healthier food is more reasonably priced because it's subsidized. But soda pop, for example, is not subsidized. "A can of pop can be $5," Gearheard says, "especially when you get into summer and supplies are dwindling. The supply ship only comes once a year." As a result, Gearheard and her husband, along with many families in Clyde River and throughout the Arctic, have come to rely on a commercial air freight service called Food Mail that is based in Quebec.[7] "They basically do your shopping for you. You send a list by e-mail or fax and they shop it all up and get it together and package it," Gearheard says. Food Mail has a contract with the Canadian government, which subsidizes the service. "So you send in your list on Monday and it comes on Thursday," explains Gearheard. "You can get all kinds of fresh vegetables, essentially everything you can get down south."

"In the past, prior to settlement, people didn't have a choice. It wasn't a question of whether you wanted to hunt," says Tristan Pearce, a graduate student finishing his PhD at the University of Guelph in Ontario, Canada. "It was a question of whether you wanted to eat." But now the Inuit are faced with choosing between their traditional lifestyle and modern conveniences and the result is often a blend of the two. "Now it's a question of do you want to eat *country foods* like seal, musk-ox and whale or do you want to go to the store and buy processed foods."

Pearce works in the town of Ulukhaktok, formerly Holman, in the Inuvialuit Settlement Region. Ulukhaktok is a coastal community of approximately 430 people located on the west coast of Victoria Island in the Northwest Territories. It evolved as a permanent settlement starting in 1939, with the establishment of a Hudson's Bay Company (HBC) trading post and a Roman Catholic mission near the location of the current settlement. Throughout the 1940s and 1950s, the regional population continued to live in isolated hunting and trapping camps and came to Ulukhaktok several times a year to trade furs and socialize. The federal government shipped three housing units to Ulukhaktok in 1960 and another four in 1961. In the years to follow, some families moved to Ulukhaktok permanently, but others live there seasonally. Snowmobiles, satellite television, Christian churches, and a wage economy all brought profound social change to this group, traditionally known as the Copper Inuit. The Copper Inuit speak Inuinnaqtun, and Western Inuit from the Mackenzie Delta region who also live in Ulukhaktok speak Inuvialuktun. But now English is the dominant language for younger people.

Despite this modernization, the Inuit are the ultimate survivors, technologically sophisticated, extremely adaptable, and yet traditional at their core. Thanks to research by Gearheard, Pearce, and others, traditional Inuit knowledge of climate and weather is gaining more attention and much-deserved respect from within the broader scientific community.

For a long time, traditional knowledge of weather and climate was classified as anecdotal by western science, *anecdotal* being a code word for *unscientific*. Yet as we learn more about traditional knowledge of climate, and collaborate with indigenous peoples, scientists have gained a deeper understanding of the interconnections within our climate system. It turns out that old-fashioned firsthand observations have a very important place in the high-tech world of modern science.

A perfect example of how traditional knowledge works its way into Western science involves potato farmers in the Andes and El Niño. Climate scientists and anthropologists had long heard stories about traditional forecasts developed by Indian farmers in the Andes Mountains of Peru and Bolivia, dating back to the late sixteenth century. The anecdote went like this. Potato farmers would meet in small groups at each winter solstice in late June (the southern hemisphere winter) to discuss the planting date for the potato crop. Then, in total darkness, during the longest and coldest nights of the year, they would climb to a mountaintop in order to see the Pleiades, a star cluster in the constellation Taurus. The Pleiades are visible low in the sky to the northeast just before dawn. In years when the Pleiades looked big and bright, the farmers would plant potatoes at the usual time. But in years when the Pleiades looked small and dim, they would expect the rains to arrive late and be sparse, and so they would postpone planting by several weeks. The farmers were using the appearance of the Pleiades to forecast the timing and quantity of precipitation during the rainy season, several months later, beginning in October and extending through March.

Mark Cane, a professor at Columbia University, and John Chiang who at the time was his graduate student and is now a professor at Berkeley, were able to find the physics underlying this traditional forecast by using modern satellite data. Cane and Chiang knew there was a strong link between El Niño and precipitation in South America. Rainfall over the Andes is lower during El Niño years. This relationship is obvious for the three months of the year with the highest rainfall: December, January, and February. More important, rainfall in October is also diminished by El Niño, and this suggests that the rainy season starts later during El Niño years. But how could El Niño alter not just the weather but also the apparent brightness of the Pleiades in June, four months before the rainy season begins?

This was where modern technology became very useful. Using satellites capable of measuring cloud cover from space, Chiang showed that there was an increase in high clouds during El Niño years. Specifically, high clouds just to the northeast of the Andean highlands increased during late June, interfering with observers' view of the Pleiades. In other words, the brightness of the Pleiades in late June indeed correlates with rainfall during the growing season for potatoes the following October through March.[8] This climate forecast is one of many that have come to the attention of scientists, and it reinforces the importance and significance of traditional knowledge.

In the Arctic, traditional forecasts always come back to snow and ice. And unlike the Andean farmers, who were focused on potato yields, the Inuit are interested in elements that affect hunting. For example, they look at the time it takes for sea ice to reach a certain thickness, at wind strength, and at the relative timing of sea ice breakup and animal migrations. "Ice is extremely important because it is essentially the highway over which the Inuit travel to hunt," says Pearce. Freeze-up generally occurs between the end of October and mid-November, and breakup usually occurs in late June or early July.[9]

Inuit travel by snow machine towing large sleds called *alliaks* in Inuinnaqtun or by dog team over the sea ice, river ice, and lake ice to reach a number of hunting areas, both on the land and on the ice. The coldest months of the year, December through March, are considered by community members to be the safest for travel, owing to the thickness and stability of the ice. During the winter, hunters in Ulukhaktok use the sea ice to hunt seals and polar bears, trap foxes, get to musk ox harvesting areas, and travel to neighboring communities. Traveling on the sea ice is inherently dangerous because of rough ice, cracks, open water leads, and storms, but hunters manage these risks by taking precautions and applying their knowledge of local ice conditions.

In the last several years, however, changes in the climate have

altered and in some cases increased the magnitude and frequency
of hazards that hunters have to deal with. In particular, some areas
of sea ice, over which harvesters are accustomed to travel, are no
longer stable, and in some instances the ice has not formed, because
of strong winds and milder temperatures. Even experienced harvesters have encountered hazards in what are thought to be safe travel
areas. Hunters are now often taking risks to travel on the sea ice
even when it is melting or thin to reach hunting areas.[10] Recently,
several hunters have been stranded, injured, or forced to take alternative travel routes, or have had to deal with lost or damaged equipment (such as snow machines breaking through the ice) as a result
of unexpected changes in weather and sea ice conditions.[11] These are
many of the same the types of issues Gearheard and her husband
had to confront during their 400-mile dogsled race. I wonder if her
mom still worries.

"Typically, before going out on the land or sea ice, hunters consult with other hunters and elders about sea ice and weather conditions. They observe the height and form of clouds, the brightness
and movement of stars, and the direction and strength of wind,
to attempt to forecast the weather in order to decide if it is safe to
travel," explains Pearce. Hunters also often consult satellite imag-
ery of the sea ice and weather forecasts available on the Internet,
merging traditional and new weather forecasting techniques. As
for navigation, compasses are of little use in the Arctic because of
its proximity to the magnetic north pole, and hunters often rely on
traditional techniques such as observing snowdrifts, memorizing
landforms, and using celestial navigation (stars) to guide them to
their destination.

"For example, when hunters travel on the sea ice, they look at
the direction and form of snowdrifts that are created by the wind.
Snowdrifts indicate the prevailing wind, which hunters use to identify the direction they are traveling. When it is dark out or there is
a blizzard, hunters can use snowdrifts to guide their dog team or
snow machine in the direction they want to go, either crossing the

snowdrifts or traveling alongside them, a traditional cruise control," Pearce says.

> When I was younger I remember that the ice freezes at the end of September, or the first week of October . . . now it freezes [in] late October, even [the] first week of November.
> —H. Ittusardjuat, Igloolik, 2004 [12]

> Long ago the cold gradually set in and the ice gets thicker. Now [there are] long spells of strong winds and the ocean can't freeze up.
> —H. Ittusardjuat, Igloolik, 2004 [13]

"I think that the clouds and the winds are two of the most common things that people use. And people talk a lot about how those have changed," Gearheard explains. "What's really interesting about the traditional forecasting is there's no general set of rules that people use," she adds. "What I've learned is that it's very individual. People might use the same indicators, but the way they use them or read them is a little bit different for each person. The cloud formations, what kind of cloud they are, what direction they're moving. And they also observe the different levels like lower clouds and upper clouds and how the clouds are formed in relation to what wind is blowing at the time." In the Arctic, where the weather is a matter of life and death, each forecast is very personal.

This traditional knowledge is passed down through the generations and combined with repeated personal experience and observations. Each generation learns how to evaluate risks, what preparations to make before going out on the ice to hunt, and what to do in an emergency. Yet although the Inuit use traditional techniques to forecast the weather, a lot of people will also turn on the television or radio to check the weather forecast before they head out on the ice. "Most hunters use a blend of both traditional and new weather forecasting and navigational techniques, with the traditional techniques still holding precedence when push comes to shove," says Pearce.

As this blend of the old and the new takes over, the younger generation loses its connection to traditional knowledge. Pearce attributes the decline partially to the southern educational system that has made its way into the Arctic. Hunting and traditional knowledge are not things teachers from the south bring with them. The less time the members of the younger generation spend hunting, the greater their dependence on wage employment, and the more they become separated from their elders. But the disconnect is also a matter of lack of "necessity." After all, who needs to learn how to predict the weather if television does it for you?

Consequently, certain skills necessary for safe and successful harvesting are not being learned; these include traditional forms of navigation, knowledge of wildlife migration patterns, and the ability to make snow shelters. Also, weaker social networks compromise the ability to cope with changing climatic conditions. Although new technology and institutions can help fill the gap, these sometimes serve to further erode traditional knowledge.

It is more dangerous [for the younger generation] because they don't know the conditions, what to avoid.

—Kautaq Joseph, Arctic Bay[14]

I think we have lost the skills so much. I mean, what would have not been dangerous for a man 50 years ago is now dangerous . . . because we have lost so many skills.

—James Ungalak, Igloolik[15]

We go to areas where we wouldn't normally go because we are assured [by the GPS] we will know where we are. . . . We [also] take more chances.

—Nick Arnatsiaq, Igloolik[16]

The dog teams know the thin ice and the thicker ice so [people] know that they can walk through thin ice. Snow-mobile doesn't say, "Alert! This is thin ice." So it's more dan-gerous [by snowmobile] than by dog team.

—Herve Paniaq, Igloolik, 2004[17]

If you don't know the traditional knowledge, you won't last very long: you will freeze to death if you don't know how to survive.

—David Kalluk, Arctic Bay[18]

"Something that I find interesting," Pearce says, "is that when I ask young hunters if they have traveled on the sea ice on their own or as the leader of a group, it's seldom that someone un-der thirty-four says yes. It's more common for someone to say that they have been out with their grandpa or other relative but not on their own.

"The sea ice has always been dangerous, but climate change has exacerbated some risks, and hazards associated with travel on the sea ice are becoming more common, such as thin and unstable ice in areas where it is expected to be thick. We're seeing very experi-enced hunters go through the ice."

But the problem is not just that the younger and the older gen-erations are not communicating; it's that some traditional ways of forecasting are beginning to fail.

The foundation of the wind has changed, it's gone. The wind will now come from any direction, any time of day. Before you could predict [the wind] but not anymore: [the wind] will be from the south and then the same day the wind shifts direction.

—David Aqiaruq, Igloolik, 2006[19]

The weather nowadays is unpredictable. You can check the five-day forecast but that doesn't mean that's the weather you're going to get.

—G. Lundie, Churchill, 2006[20]

"Inuit do recognize that there is knowledge erosion, but for weather forecasting the problem is not that they don't know how to do it anymore; many people do," Gearheard explains. "It's that their forecasting techniques no longer fit the weather that's happening now. And because forecasting the weather is literally a matter of life and death, they don't want to teach it because it might put someone in harm's way."

———

Perhaps because of what Gearheard, Pearce, and others have seen personally, many scientists believe that cultural preservation, along with housing and infrastructure improvements, is an important way to help the Inuit simultaneously tackle the issues of climate change and cultural erosion.

The Inuit still hunt narwhals, ringed seals, walrus, beluga whales, arctic char, caribou, polar bears, and a variety of migratory birds.[21] But they see changes in wildlife, too. Ringed seals, for instance, are believed to be particularly susceptible to climate change. They depend on the sea ice for pupping and snow cover in spring to build their birth lairs.[22] All these conditions will be affected by climate change. Climate change is also likely to increase harvesting pressure on seals; easily accessible year-round near most Nunavut communities, seals become "fallback" prey when hunters cannot reach hunting grounds for other species. Caribou are also believed to be susceptible to climate change.

For the Inuit, climate change means greater risk and greater uncertainty. Their own observations have indicated changes in temperature and precipitation, permafrost, coastal erosion, and ice instability.[23] For coastal communities such as Shishmaref in Alaska, perhaps the prime example of climate change in the Arctic, sea ice serves as a buffer against battering waves during severe Arctic storms. Inuit knowledge is already evolving in response to climate change. The increasing unpredictability of the weather and sea ice is becoming part of the collective social memory.

> My aunt, Mable Tooli, said [to me]: "The Earth is faster now." She was not meaning that the time is moving fast these days or that events are going faster. But she was talking about how all this weather is changing. Back in the old days they could predict the weather by observing the stars, the sky, and other events. The old people think that back then they could predict the weather pattern for a few days in advance. Not anymore! And my aunt was saying that because the weather patterns are [changing] so fast now, those predictions cannot be made anymore. The weather patterns are changing so quickly she could think the Earth is moving faster now.
>
> —Caleb Pungowiyi, 2000[24]

Scientists will tell you that climate change is happening faster in the Arctic than anywhere else on the planet. They have been tracking the big picture across the Arctic using every high-tech tool available to them, including satellites and radar. Average temperature has risen almost twice as fast across the Arctic as in the rest of the world during the past few decades. And it's not just the temperature that is moving fast. There is also a widespread melting of glaciers—and a thawing of permafrost, ground that was until now permanently frozen. Permafrost has warmed almost 3.5°F in recent decades. The warming is affecting villages' water supply, sewage systems, and infrastructure, as pipes sit aboveground and are vulnerable to any shifts associated with permafrost. The tree line, another boundary distinguishing the Arctic from points south, is creeping north. And snow cover has decreased about 10 percent during the past thirty years. Winter temperature in Alaska and western Canada has increased about 5°F to 7°F during the past fifty years. Winter is becoming shorter and warmer, and that is not good news if you're a winter person.

There also is the sea ice. The physics of much of this fast change goes back to the ice. First, as Arctic snow and ice melt, darker land and ocean surfaces open up. They absorb a lot more of the sun's energy, and therefore the temperature spikes up faster. This is a process called *Arctic amplification*. The temperature is amplified by the loss of ice. And so it follows that one of the coldest places on the planet is warming up fastest.

Scientists have been carefully tracking the extent of sea ice using satellites since 1978. The extent of Arctic sea ice has a natural cycle: it grows and shrinks with the seasons. Twenty-five years ago, the seasonal range usually peaked in March at about 6 million square miles and shrank to about 3 million square miles at the end of the summer melt season in September.

As the climate grows warmer, the summer melt season lengthens and intensifies, resulting in less sea ice at summer's end. With nearly twenty-four hours of sunlight hitting areas of open water, the ocean heats up more and the open area grows larger. Summer eats into autumn and winter, and sea ice formation, critical for insulating the warmer ocean from the cooler atmosphere, is delayed. Less sea ice at the end of summer pushes more heat into the atmosphere in autumn. The seasons are out of sync. It is only once the sea ice forms that heat exchange is finally capped. With continued summer ice loss and more ocean heat gain, Arctic amplification will eventually eat its way into winter.

Aside from Arctic amplification, there are other reasons the Arctic is seeing such rapid change. One reason has to do with the fact that the air in the Arctic holds less moisture (because it's so cold). Because of this lack of moisture, a greater fraction of the energy that comes from increasing concentrations of greenhouse gases can go directly into warming the atmosphere. In places that are more humid, such as the tropics, the energy is split between warming the air and evaporating the moisture. In other words, in the Arctic, $CO_2$ has one less job to do. It can focus strictly on warming things up.

There's been a lot of talk within the scientific community about when the Arctic might become ice-free in summer. Since the start of the modern satellite era in late 1978, the extent of Arctic ice has shown a downward trend across all months, with the largest decrease occurring in September. The decline in September is about 12 percent per decade. And starting in about 2002, the pace of melt seems to have picked up. The extreme seasonal minima of 2007 and 2008 reinforce this tendency. Since 2002, scientists have watched in what can best be described as shock as the September decline has continued. According to analyses by the U.S. National Snow and Ice Data Center, where Gearheard works, the September minimum was set in 2007, but 2008 entered the record books as the second-lowest of the satellite era, and probably the second-lowest of at least a century.

Scientists also track the overall thickness of the ice, because this has significant implications for what the Arctic will look like in the future. Ice thickness also has its own rhythm. Back in the 1980s, when the system was still stable and running like clockwork, about 40 percent of the ice in April would consist of young, fairly thin ice that had formed the previous autumn and winter. The other 60 percent consisted of thicker ice that had survived one or more melt seasons. Generally, the older the ice, the more melt seasons it has survived, and the thicker it is. This distribution would change with the seasons. Most of the young, thin ice melted to form open water areas while some of the older ice thinned in summer, and only a bit of it actually melted. The ice that survived the summer thickened again the next autumn and winter. Over the course of each year, ice growth exceeded melt. That doesn't happen anymore.

In recent years, old ice has been harder to come by. The thickness distribution has shifted in favor of the young, thin ice. Because the ice is thinner at the start of the melt season, open water areas develop earlier than before and become more extensive through the summer. As a result, the amplification that boosts the melting is even stronger. The past few years have seen the distribution shift to even thinner spring ice, resulting in even larger open water areas absorbing solar radiation, and an even stronger feedback. The Inuit are seeing it, too.

"People here in Clyde River will say that they've lost three to four weeks of sea ice already," says Gearheard. "The sea ice is so important in terms of hunting and travel. But people so far have been able to cope with it. But it's things like the weather changing and losing their ability to forecast that is really hitting home. That is a significant impact," Gearheard adds.

"You can already see some of the ways the Inuit are adapting," says Pearce. "For example, we're already seeing people invest in larger boats with larger motors because the open water season has lengthened. Some people see a real opportunity to having a boat because they can travel to fall and spring hunting grounds despite the absence of sea ice."

Inuit have no choice but to adapt. And Gearheard and Pearce feel strongly that involving Inuit communities in the research process will help them develop the necessary adaptation strategies. With that in mind, Gearheard and some other community members in Clyde River developed the Igliniit project. The project is part of the Inuit Sea Ice Use and Occupancy Project. An International Polar Year project, Igliniit brings together Inuit knowledge with engineering and cutting-edge technology. In Inuktitut, *igliniit* refers to the trails that are routinely traveled by hunters and other members of the community.

The location, use, condition, and changes in these *igliniit* over time and space can help both community members and other researchers learn a great deal about the Arctic. Engineers and Inuit hunters have come together to design a new, integrated GPS system that can be mounted on snowmobiles. The GPS automatically logs the location of the snow machine every thirty seconds, providing geo-referenced waypoints that can later be mapped to produce the travelers' routes on a map. In addition to tracking routes, the Igliniit system logs weather conditions (temperature, humidity, pressure, etc.) and the observations of hunters (animals, sea ice features, hazards, place-names, etc.) through a customized computer screen that is in Inuktitut and has a user-friendly icon interface.

The hunters carry digital cameras that allow them to take pictures of conditions as well as make videos. The data and images can then be downloaded and turned into maps. These maps show the routes of individual snowmobiles, along with the geo-referenced observations of the hunters and weather conditions. When you overlay maps of different hunters over time and space, you end up with a valuable picture of meteorological data and Inuit land, ice, and resource use. At the moment, the Igliniit system is being tested on snowmobiles and dog teams, but Gearheard hopes it will eventually be made available to boats and all-terrain vehicles.

Gearheard and her team hope the Igliniit system will provide useful information for communities like Clyde River. Hunters can print out their own maps to keep a record of their travel routes and hunting spots. And collectively, the community can use the maps to see where its members have had the most hunting success, changes in animal populations, changes in snow conditions, connections between weather conditions and travel conditions, and locations of hazards.

"In traditional Inuit society people share food," explains Pearce. "It can be community to community as well as household to household. For example, when Ulukhaktok has limited access to caribou, a nearby community in Nunavut will help them out and send caribou. Or if Sachs Harbor has geese or Tuktoyaktuk has beluga whale and Ulukhaktok doesn't, they will send geese and *muktuk* (whale meat). Whereas Ulukhaktok might have a really good char run that year, so they will send char."

These maps might also help preserve traditional values. In addition, community leaders can use the maps for matters related to planning or negotiating land use. The weather information is also extremely valuable, as weather data are spotty in the Arctic, restricted to a limited number of weather stations. Last, Igliniit has the potential to serve a role in search and rescue operations. Igliniit maps can be used to keep up-to-date records of sea ice, land, and water hazards logged by hunters, and these maps can be easily shared with other hunters. Gearheard's team is currently looking into the possibility of incorporating live tracking or personal locator beacons (PLBs) into the Igliniit system, so that anyone carrying an Igliniit unit can send out a warning signal in order to be tracked quickly.

"You can look at climate change in isolation, but I think it would be a real disappointment," says Pearce. That is exactly why he, Gearheard, and others are working to connect climate change to other issues facing Arctic communities. By connecting climate change to issues like education, sustainable development, and alleviating poverty, you can reduce vulnerability across the board. When you find ways to strengthen Inuit communities today, you can also strengthen Inuit communities in the future.

"I'm always hearing the words 'in the future' or 'future projections,' " says Pearce. "But when I am in the Arctic, I am constantly struck by the fact that the future has arrived, and Inuit are living it. It's happening now." Under the scenario of the Intergovernmental Panel on Climate Change (IPCC) for "business as usual" greenhouse gas emissions, simulations from the current generation of coupled global climate models indicate that the Arctic will warm between 7°F and 13°F over the next 100 years. And the Arctic Ocean could become seasonally ice-free, or nearly so, sometime after 2040.

Whereas most climate model simulations show the September loss of sea ice increasing in the coming decades, this acceleration seems to have already begun. In other words, when you compare the present with future projections from climate models, we are already on the fast track of climate change. The Earth, as the Inuit say, is faster now.

I asked Gearheard what she imagined the Inuit might be doing fifty years from now if all the summer sea ice disappears. "I guess when I think about the future, I just think about the people that I know right now and try to imagine what they would be doing. And I can't imagine that the people I know would just stop hunting or stop going out on the land. Every person that I know who is active in that way would still be doing it. They might be hunting something different. They might be hunting somewhere else. They might be selling their ski-doo and buying a boat. But no matter what happens, they would still be Inuit and they would still be out there hunting."

This raises a question: what happens to the Inuit if the snow and ice go away? "That's a really big question," says Gearheard. But she sees the connection to snow and ice and tradition even in very young children in Clyde River. "If you ask them to draw a picture of their family, you'll very often get a picture of their family on a dogsled traveling over the ice. Even if they don't have one," Gearheard says. "The snow, the ice, it's all wrapped up in there."

"The truth is," says Pearce, "in the Arctic, people have a strong connection with their environment. This is their home; this is where their life is. They are going to adapt to new conditions regardless if any action is taken against climate change. They are not going anywhere. They make that very clear. If you ask an Inuit in Ulukhaktok what their future aspirations are, they will tell you that they plan on staying in the Arctic. End of story." Sea ice or no sea ice. I guess that also applies to the rest of us.

# 9

# THE ARCTIC, PART TWO: GREENLAND

Erik the Red may have deserved to go to hell, but instead he went to Greenland. In A.D. 982, after being banished from Norway and then Iceland for murder, the infamous Viking explorer headed west. Only 500 miles from the shores of Iceland, he discovered a beautiful island with rich pastureland tucked into deep fjords overflowing with crystal blue water. When his term of banishment expired three years later, he returned to Iceland, in the hope of finding a few brave souls willing to join him and settle this little slice of heaven he had discovered. Erik the Red christened his new home *Greenland*.

It has been suggested that the naming of Greenland is a very early example of bait-and-switch advertising—a deceitful way to get warm bodies to a cold place. The right name may be all it takes to sell people on a dream of a better life and a brighter future. On the other hand, although Erik the Red seems to have been a murderer, he wasn't necessarily a liar. You might say that in naming Greenland, he told a white lie. At the time, Greenland was actually much greener than it is today. From about A.D. 800 to 1300, the Medieval Warm Period, Greenland's climate was much milder, and the southern part of Greenland, where Erik the Red settled, did indeed have meadows lush with grass, willows, and wild berries.

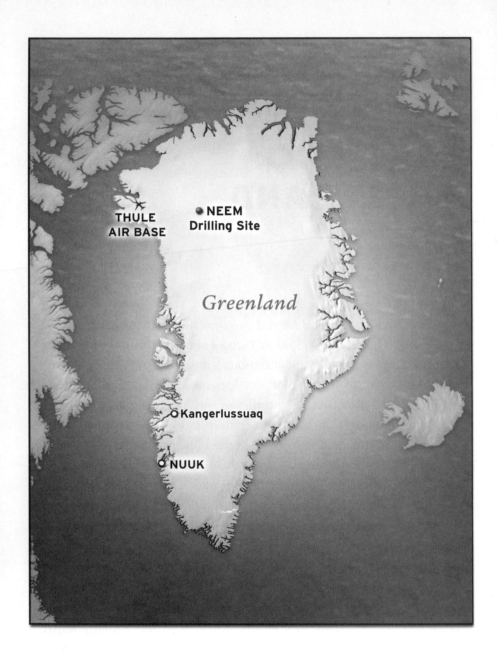

THULE
AIR BASE

● NEEM
Drilling Site

*Greenland*

○ Kangerlussuaq

○ NUUK

In any case, Erik the Red must have been a good pitchman, because in 985 he led a fleet of twenty-five Viking longships to settle two new colonies on Greenland. During the next ten years, as the news of free pastureland traveled back to Iceland, three more ships carrying hopeful settlers set sail for Greenland. And by the year 1000, virtually all the land suitable for farms in the Western and Eastern settlements of Greenland had been claimed. About 1,000 people lived at the Western Settlement and 4,000 people at the Eastern Settlement, which, despite its name, was located about 300 miles to the south.[1] Erik the Red had successfully converted 5,000 Icelanders into Greenlanders, but he certainly hadn't led them to the promised land.

Over time, Erik the Red's original white lie grew whiter. Summers were becoming shorter and cooler, and winters were downright frigid—even by Viking standards, which were quite harsh. This limited the amount of time the cattle, sheep, goats, and horses could be kept outside to pasture and increased the need for winter fodder. As the temperature dropped, the amount of sea ice increased. The sea ice became a frozen barrier, making it more and more difficult for ships to pass. As a result, trade and communication with Europe and Scandinavia were choked off and Greenland became increasingly isolated.

The cooler climate also brought the Inuit down from the north and into more regular contact with the Norse colony. Their relationship might have served as an impetus for change, pushing the Viking settlers to find new ways to deal with the cold; but in fact it only brought more problems. The little archaeological evidence that exists suggests that there was violence between the two groups. The changing climate had ushered in a period now known as the Little Ice Age. And after 500 years of settlement, the Viking colony, unable to adapt to the cooler conditions and unwilling to supplement Scandinavian tradition with Inuit coping strategies, eventually collapsed. The last written record of the Norse Greenlanders comes from a marriage in the church of Hvalsey in 1408. The church still stands today.

In May 1721, Hans Egede, a thirty-five-year-old Lutheran missionary, received permission from Frederick IV of Denmark to search for Erik the Red's lost colony. No word had come out of Greenland for more than 300 years, and Egede feared that the Viking colony was lost or, perhaps worse, that the colonists had lost their faith. And so Egede and his wife set sail from Bergen, Norway, and headed for Greenland, where they intended to set up a mission. Upon their arrival, Egede found no Norse survivors. He did, however, find the Inuit. And so he started his mission among them.

Egede, called the apostle of the Eskimos, spent fifteen years in Greenland. He studied the Inuit language and tried his hand at translating Christian texts, a task that required the ability to adapt the text in such a way that his teachings would resonate with the Inuit experience. For instance, the Inuit did not eat bread—a fact that made the Lord's Prayer rather cryptic. Egede made one small but critical adjustment and wrote, "Give us this day our daily harbor seal." It seems that Egede decided to leave the concept of hell unaltered, despite the cold climate. The Inuit learned about a very hot place where sinners were sent for eternity.

Today, new settlers are traveling to Greenland in search of the promised land. But they come in corporate jets rather than Viking longships, and this time the Inuit are happy to see them. Greenland figuratively and (as I will explain later) literally is on the rise.

Gold and diamond prospectors are heading to the southern part of Greenland. Alcoa, the U.S. aluminum giant, is preparing to build a smelter powered by hydroelectric energy in Maniitsoq, on Greenland's west coast. In addition to precious metals and diamonds, Greenland also has oil and natural gas. The United States Geological Survey (USGS) recently assessed the area north of the Arctic Circle and concluded that about 30 percent of the world's undiscovered gas and 13 percent of the undiscovered oil may be found there, mostly offshore and under less than 550 yards of water.[2] The USGS estimates that the Arctic could hold 90 billion barrels of oil and 1,670 trillion cubic feet of gas, much of it off Greenland. East Greenland alone is estimated to contain 8.9 billion barrels of oil. Exxon Mobil and Chevron are already increasing their exploration. But there's a catch.

Greenland is hidden under a 1.6-mile-thick layer of ice known as the Greenland Ice Sheet (GIS). And with more than 80 percent of the island essentially on ice, much of Greenland's potential wealth is more or less wishful thinking. Greenland may not be the promised land quite yet, but it is a land full of promises.

A Danish protectorate since 1721, Greenland has long sought to sever ties with its benevolent colonizer, a colonizer that supports the island with an annual grant of $600 million. On June 21, 2009, Greenland came one step closer to eventual independence from Denmark by voting in a new era of self-governance. Under the new self-government agreement, Greenland will keep half of all proceeds from oil and mineral finds. At first it will also continue to receive the $600 million annual grant from Denmark, but as petroleum and mineral revenues increase, the grant will be reduced—and it will continue to be reduced until it hits zero. Greenland can choose to secede from Denmark anytime along the way. As the ice melts, the money will rush in—or so the thinking goes.

If a changing climate helps speed that process up, then many Greenlanders say bring it on. For the people of Greenland, 90 percent of whom are indigenous Inuit, the question of how quickly Greenland's ice will melt is not merely an abstraction; it represents freedom. However, there's another catch. If the GIS melted, exposing all the riches beneath it, it would also raise sea level worldwide by 23 feet.[3] This is why the loss of the GIS represents one of the worst-case scenarios with regard to global warming.

There is a rather straightforward accounting system as it applies to global warming in Greenland. There are pros and cons, pluses and minuses, gains and losses. The larger question is: how will these

add up? Warmer winters are already making life tough for traditional communities that hunt and rely on predictable sea ice (see
Chapter 8). Hunters who use the sea ice for hunting and travel have
found themselves idle when the ice fails to form and the whales,
seals, and birds they hunt shift their migratory routes. Melting
permafrost is buckling roads and airport runways, raising costs for
the mining companies that are seeking aluminum, diamonds, gold,
zinc, and more. But the warmer weather also encourages tourism,
and the loss of ice means that the ship transport season in the Arctic
is easier and longer. Fishermen report a rise in some fish stocks,
including cod and halibut, as a result of the warm-water currents
that now flood into Disko Bay. Shops in the island's capital, Nuuk,
have even begun to offer homegrown potatoes and broccoli—crops
you don't necessarily associate with Greenland. Whether you think
in terms of dollars, temperature, glaciers, or even broccoli, there are
plenty of people who like what they see.[4] Ultimately, though, the
issue comes back to the ice. And you can argue that whatever factors control the amount of ice control Greenland's destiny. That is
where the scientists come in.

———

Scientists have been trying to understand Greenland's ice for a long
time. There is no doubt that Greenland fluctuates between warmer
and cooler, wetter and drier, greener and whiter. Hans Egede
Saabye—who was the grandson of Hans Egede and was also a
missionary—first noticed this.[5] Saabye evidently had a keen eye for
changes in weather and climate. He was an *observationalist* in the
classic sense of the word, collecting data and monitoring change.
Saabye wrote down these observations in his diary, and one of his
comments was particularly perceptive: "In Greenland, all winters
are severe, yet they are not alike. . . . When the winter in Denmark
was severe, as we perceive it, the winter in Greenland in its manner
was mild, and conversely." Scientists now understand that Saabye

was describing an important atmospheric pattern called the North Atlantic Oscillation (NAO).

The NAO is a phenomenon that affects weather and climate from North America to Siberia and from the Arctic to the equator.[6] It is a dominant mode of natural climate variability; and as with El Niño, scientists are working to develop a long-range seasonal forecast for it. The main feature of the NAO is a seesaw of atmospheric pressure between a persistent high-pressure cell over the Azores and an equally persistent low-pressure cell over Iceland. The NAO index, which swings between positive and negative, represents a measure of the relative strength of these two pressure systems. Depending on the phase, the NAO can bring large changes in surface air temperature, winds, storminess, and precipitation.[7] The NAO is most pronounced during the winter, and that is why the connection to the ice exists. In the case of Greenland, it appears that snow falling in the center of the island is linked to a negative NAO index.[8] But the NAO is just one of many factors affecting Greenland's ice; that is why ice sheets are a very complicated matter.

When scientists talk about the GIS, they are talking about *mass balance*. This concept implies that the amount of snow falling in the middle of Greenland is balanced by the amount melting along the sides. During the early 1990s, the GIS was in the Zen-like state of mass balance.[9] Today, scientists are talking about *mass loss* because all signs point to the fact that the GIS is melting much faster than it's growing.[10] Mass loss is caused by a combination of two factors. There is *melt*, which can result from an increase in temperature and is caused by a variety of factors. There is also a process called *calving*. Calving, a much slower process than melting, takes place when glaciers flow into the sea and eventually break away from the coast. A recent study conducted with the Ice, Cloud, and Land Elevation Satellite (IceSat)—a NASA satellite that uses lasers to calculate change in elevation—found that the glaciers are indeed speeding up where they flow into the sea.[11] Scientists think that the most likely cause of faster glacier flow is warm ocean currents reaching the coast and melting the glacier front.[12] But calving is so poorly understood that it remains one of the most unpredictable components of future rises in the sea level.

Since 2002, Greenland has come under the watchful eye of another NASA satellite mission: the Gravity Recovery and Climate Experiment, or GRACE, which some scientists call *amazing GRACE*. This mission doesn't see continents or oceans so much as it sees gravity. Since its launch in 2002, GRACE has been acquiring ultraprecise measurements of Greenland's mass loss. Scott Luthcke, a geophysicist at NASA's Goddard Space Flight Center in Greenbelt, Maryland, describes GRACE like this. "Imagine you caught a big fish and you wanted to weigh it. One way you could do it is you could go out and buy a scale. But you could also just use a spring," Luthcke begins. "The distance that the spring stretches when you hang the fish from it is a representation of how much the fish weighs. That's pretty much what GRACE is doing." Of course, GRACE, which is able to measure changes as small as the width of a human hair, is doing it (as noted above) with high precision, using microwaves that essentially act as the spring to measure the mass of the fish—the GIS. Technically, GRACE, two identical satellites separated by a distance of 137 miles, is doing it from a polar orbit 310 miles above the Earth's surface.

The beauty, or perhaps the amazing part, of GRACE is that every month since 2003 it has flown over the GIS, keeping track of its comings and goings, its growing and melting. When you fly over the same place again and again, you can measure how much it's pulled apart and compressed together and how much the mass on the ground has changed. Unfortunately, in the case of the GIS, the fish is slowly disappearing.

Luthcke is an observationalist in the purest sense. He just happens to be working with some of the most sophisticated measuring devices ever built. Data from GRACE shows that Greenland lost about 200 billion tons of water per year from July 2003 to

July 2008. That works out to the equivalent of Lake Erie draining into the ocean every two years. Needless to say Greenland is losing weight.

This enormous weight loss is a result of melting and thinning at the margins along Greenland's coast[13] as well as a surging (or acceleration) of outlet glaciers into the sea.[14] One example of an outlet glacier is the Jakobshavn Isbrae glacier.[15] Jakobshavn Isbrae is Greenland's largest outlet glacier, draining about 6.5 percent of the GIS area. It has been surveyed repeatedly since 1991 and has been accelerating since the mid-1990s. That said, this tremendous crumbling and melting along Greenland's coastal margins is partly compensated for by some mass gain in the interior of the island, a gain that is controlled in part by the North Atlantic Oscillation.[16] Even so, the GIS has shrunk so much in recent years that the underlying bedrock, like a ship emptying of its cargo, is lurching up at a rate of about 1.6 inches each year. Greenland is indeed rising.

J. P. Steffensen, a scientist at the Center for Ice and Climate at the University of Copenhagen's Niels Bohr Institute, agrees that an amount of meltwater equivalent to Lake Erie every two years is pretty impressive, but it's virtually nothing compared with what he's seen happen in his ice cores. "We've seen climate changes that would have wiped off life, changes that are just mind-boggling fast," says Steffensen. "We literally see the ice age ending from one year to the next," Steffensen says. "It's pretty scary."

Steffensen reads ice the way most of us read a book. And there is one chapter that worries him. "The Earth has always had climate changes. The problem is that right now we have a climate change which is permanent," he explains. "Even though nature has done it by itself before, we should not put ourselves into the position of kicking the climate system. That's my little worry."

Steffensen is the field operation manager for the North Greenland Eemian Ice Drilling Project (NEEM). The NEEM camp, located in the far north of Greenland (see map), brings together scientists from fourteen nations including the United States and is aimed at retrieving a core of solid ice 1.6 miles long. The goal is to unlock the climate history trapped inside tiny air bubbles from the ancient atmosphere. You could say Steffensen is old school. He prefers data to models, hardware to software, and ice cores to satellites. "There's a much-admired saying in our community," Steffensen says with a smile: "There's no substitute for data." That is where the ice cores come in. "I mean, when you do climate models, you have to realize that the models do not include the things you don't know. With ice core data, we can see things happening in the climate system that the models so far have never been able to capture. We reveal the climate history," Steffensen says.

He gives an example of two rapid warming periods he sees in the ice that constitute a profound climate shift: one at 14,700 and one at 11,700 years ago.[17] It's thought that the shift must have come from the atmosphere, which is far more nimble than the slowly churning oceans. If the atmosphere is capable of suddenly flipping into a new pattern, then this new pattern could have contributed to the rapid warming of the entire northern hemisphere.[18] Taken together, these two pulses of rapid warmth pushed the Earth's climate out of the last ice age and into the Holocene, our current warm period. The part that scares Steffensen is that during the ice age, the Earth's climate was far more unstable than it has been of late. The Holocene, in addition to being warm, is known for its rather remarkable stability. In this regard, observations of the real-world climate system and simulations from climate models have yet to overlap.

Although climate models help scientists better understand the complex mechanisms that create rapid climate shifts, the models can't seem to actually make such shifts happen. Typically, when a climate model tries to simulate the abrupt natural climatic shifts of the distant past, those shifts end up taking more than 100 years to occur—which is hardly abrupt.[19] This suggests that some aspects of the physics still need to be worked out. And Steffensen is hoping that data from the NEEM ice core can be of some use. This specific ice core will allow scientists to see the climate of Greenland over the past 130,000 years, and isolate a fascinating but poorly documented interglacial period known as the Eemian. During the Eemian, Greenland's temperature was about 5°F to 9°F warmer than it is today, so this period is a meaningful analogue for future climate.[20] Global sea level rise during the Eemian is also a matter of great interest, as it's likely that the sea level was between 13 and 20 feet higher. Steffensen and his colleagues at NEEM are looking at the Eemian to learn how quickly the ice covering Greenland might melt when the climate is that much warmer.

The NEEM project officially began in 2007, when Steffensen and some colleagues dragged equipment from the previous drill site—a camp called NGRIP—over to the NEEM drill site. Steffensen has carefully selected this new drill site on the basis of radar profiling of the internal ice layers and the bedrock topography. If an ice core really is like a book, then in this case the book is *War and Peace*, and of the 1.6 miles of ice each chapter or year of climate history during the Eemian is about one-third inch long. Ice cores, like tree rings, allow you to reconstruct climate on an annual basis. With each passing year, snow falling on central Greenland lays down a distinct layer, trapping bubbles of atmospheric gas, dust, and other impurities and gradually compacting into ice that captures an ancient climate record stretching back hundreds of thousands of years.

"Ice cores serve as a remarkable archive of past climate and atmosphere because of the bubbles of air that are trapped in the ice," explains Jeff Severinghaus, a climate scientist at the Scripps Institution of Oceanography in La Jolla, California. "And the beautiful thing about an ice core is that it has all of these different indicators: atmosphere composition, temperature, mean ocean temperature, dust." The oldest ice cores now go back about 800,000 years, and scientists are optimistic about pushing this method back even farther. Many of the scientists are involved in International Partnerships in Ice Core Sciences, which is aiming to retrieve a 1.5-million-year-old ice core. "And what's so remarkable is that you can still answer questions down to the year," says Severinghaus. "It's really like pull-ing back the veil."

What worries many scientists is that behind the veil they might find a threshold or a tipping point past which the GIS becomes perpetually out of balance, unstable, locked in a persistent state of wasting away.[21] Right now, the temperature range assigned to that tipping point is large because there is a high degree of uncertainty. The IPCC puts the temperature increase at somewhere between 3°F and 8°F above the long-term average. Given the 1.3°F of warm-ing we've already put into the system, it's a temperature range that could easily be in the cards during this century.

A fundamental problem is that existing models of ice sheets are unable to explain the speed of the recent changes in the GIS that GRACE and IceSat are observing. In other words, the models cannot reproduce the data. Scientists such as Scott Luthcke are seeing things happen in Greenland right now that, technically, the models don't show as happening for another thirty years.[22] Even if some temperature threshold is passed, the IPCC gives a 1,000-year timescale for a total collapse of the GIS. But, given the inability of current models to simulate the rapid disappearance of continental ice right now, let alone at the end of the last ice age, a lower limit of 300 years is conceivable.[23]

———

I met up with Steffensen, Severinghaus, and other scientists from the NEEM project in Kangerlussuaq, a former Cold War outpost for the U.S. Army and now the site of Greenland's major interna-tional airport. Steffensen, who is Danish, has spent much of his life studying Greenland's ice. "I totaled it up," he says, taking a puff on his pipe. "I'm fifty-two years old now and I've been to Greenland twenty-three times. That works out to spending almost six years on the ice. I guess you could say the ice went straight into my belly and it stayed there."

Kangerlussuaq also serves as the staging ground for the NEEM scientists who are flying north to the ice camp. This time Steffensen will stay behind to manage the logistics and make sure the ice cores get safely onto the bright red Greenland Airlines jets bound for Copenhagen. But he is no stranger to life in a remote field camp. His first season on the ice was in 1980. "It was a marriage for life," he says solemnly. That statement turns out be more true than I realized. Steffensen's wife is a fellow scientist, Dorthe Dahl-Jensen, also a professor at the University of Copenhagen and the project leader at the NEEM drill site. She shares with Steffensen the difficult task of coordinating the drill teams and the scientists, as well as making sure the ice cores get safely from the drill camp to laboratories around the world. "But my main interest is really the ice," she says. "So, I normally find the time every day to go down into the science trench and work with the core samples, because that's really my heart," she says with a shy smile.

If the NEEM project is successful, it will be the first complete record of the Eemian. None of the former deep ice cores from Greenland contain complete and undisturbed layers from this warm period, because the layers had either melted or been disturbed by ice flow close to the bedrock. "The last several ice cores in Greenland tried to get this interglacial period but didn't quite succeed," explains Jeff Severinghaus. "NEEM is really trying to get a record of the last time that the Earth was warmer than today," he explains. "So the Eemian is really an analogue of what our future looks like under global warming. It's a very, very realistic scenario for what we may experience in the next 100 to 200 years."

"We know that even though it was warmer in Greenland, it wasn't warm enough for the whole Greenland Ice Sheet to disintegrate," explains Dahl-Jensen. "And that's something that is pretty hotly debated." Specifically, this is the question of the tipping point, or how much warming we would need in the future before the GIS would totally disappear. Scientists already know that shrinkage of the GIS during the Eemian contributed an estimated 6 to 10 feet to the global rise in sea level, although a widespread ice cap still remained over portions of Greenland.[24] "Our earlier results tell us that during the Eemian, the Greenland Ice Sheet was about 30 percent smaller. And that tells us that roughly 3 to 7 feet of global sea level rise came from the Greenland Ice Sheet alone," Dahl-Jensen says. "We can also see that when the climate is warmer, it is also very stable. And that's another big debate. If we aggressively warm the climate, will it shift back into an unstable regime?"

If the weight of 130,000 years of climate history isn't enough pressure on these delicate layers of ice, the hopes of the climate science community are also bearing down. "I feel a sense of awe when I am able to peer into the deep, deep past time," explains Severinghaus. "It's very hard to put into words, but it's really quite a sense of excitement and wonder and mystery." And perhaps just a touch of dread.

The NEEM field camp accommodates about thirty researchers and technicians from May to August. "It's kind of like a frontier outpost up here," says Vasilii Petrenko, a scientist at the University of Colorado. "It's a very simple life. We're working for about fifteen hours a day most days. But we take breaks. We come back for lunch; we come back to warm up sometimes, get some tea and cookies. And everybody gathers in camp for dinner." The food at NEEM gets universal raves. "I can honestly say that I eat better here than I do at home. It's a very calorie-rich diet, but you need it here. Once you've spent a couple of days in the science trench, your body starts adjusting to the cold, and producing more heat, so our calorie requirements just skyrocket. I probably eat twice as much here as I do normally at home."

The field camp may look like a frontier outpost on the surface, but the scientists are engaged in some very sophisticated climate research in the underground science trench dug from the snow 30 feet below the surface where the ice core drilling and initial processing is done.

Developing a climate history involves reading many types of measurements from the wide variety of particles that get trapped.

Oxygen isotopes are a proxy for local temperature; excess deuterium is a proxy for ocean surface temperature; dust and calcium originate from low-latitude Asian deserts; sodium indicates marine sea salt. Impurities in the ice reflect the impurity load of the atmosphere of the past, and gas bubbles trapped between the snow crystals contain samples of the actual atmosphere, reflecting the amount of greenhouse gases such as carbon dioxide. The crystal structure of ice and the content of biological material also provide information about past climatic conditions. Volcanic eruptions can be used to date the ice. A peak of volcanic dust in the ice core allows you to match it to a volcanic eruption and have an independent estimate of age.

"All of these kinds of different indicators are on exactly the same timescale, so you can really make detailed comparisons between one indicator and another," explains Severinghaus. "Carbon dioxide is a strong greenhouse gas. So, one of the things that we see in the ice cores is a strong correlation between carbon dioxide levels and temperatures. So at times of warm temperatures, carbon dioxide is high; at times of cold temperatures, carbon dioxide is low—which reinforces what science has been showing recently: that carbon dioxide does cause warming. During the Eemian, carbon dioxide was definitely lower than today, a lot lower than today," says Petrenko.

"It suggests that today's carbon dioxide levels are entering the danger zone, basically. The Earth hasn't seen these levels of $CO_2$ for millions of years, which means that we are headed for a climate that is beyond anything that the ice cores can show," Petrenko continues. With regard to the future and the still unwritten chapters of climate history, past data can take you only so far. "The Earth is not a system that you should do experiments on," says Severinghaus. "Better to look at the experiments that Mother Nature did on her own, in the past, and study the results of those experiments. That's why looking at ice cores is incredibly valuable," he adds.

Back in Kangerlussuaq, I meet up with Steffensen before the long flight home. He reflects on the work of the NEEM researchers and on the looming issue of climate change. "We have to get used to this word *change*," he says. "That's why we have a past, why we have a future—time is flowing forward. We should never strive to re-create the past. That's impossible, because nothing will ever be as it has been. So my point is we should look forward with hope. But," he adds after a moment, "we should never forget that nature can also turn dreadful." Just ask Erik the Red.

## The Arctic: The Forty-Year Forecast—Ice Melt, Mineral Resources, and a Hospitable Arctic Circle

| CLYDE RIVER, NUNAVUT | TODAY | | 2050 | | 2090 | |
|---|---|---|---|---|---|---|
| Emissions Scenario | JAN. | JULY | JAN. | JULY | JAN. | JULY |
| Higher Emissions | 18.6 | 8.9 | −12.6 | 40.6 | −3.7 | 42.7 |
| Lower Emissions | | | −13.9 | 40.2 | −10.4 | 41.1 |

| KANGERLUSSUAQ, GREENLAND | TODAY | | 2050 | | 2090 | |
|---|---|---|---|---|---|---|
| Emissions Scenario | JAN. | JULY | JAN. | JULY | JAN. | JULY |
| A2 | 0.6 | 9 | 3.2 | 51.4 | 8.5 | 54.2 |
| B1 | | | 3.2 | 51 | 5.3 | 51.8 |

## Forecast
## June 2011

The northern coastline of Alaska was changing. Midway between Point Barrow and Prudhoe Bay, a stretch of coastal cliffs as long as a football field was being devoured by the ocean every three years. As bigger and bigger waves pounded away at the shoreline and warm seawater chipped away at their base, the 12-foot-high bluffs lining Prudhoe Bay toppled into the Beaufort Sea—more than 30 feet disappeared every year.[25] Up here, climate change was a triple threat, involving warmer oceans, stronger waves, and shrinking sea ice.

Arctic sea ice was now declining at a rate of almost 12 percent per decade. And to make matters worse, less than 20 percent of the ice cover was more than two years old—the lowest amount ever recorded since satellite measurements began. The young ice, so thin and fragile, didn't stand a chance.

Despite the dramatic changes taking place across the Arctic, it wasn't the satellite data or the retreat of sea ice that transformed many skeptics into believers. It was the Russians. When the Russians planted a titanium flag at the bottom of the Arctic Ocean, in order to lay claim to the possibly vast oil and mineral deposits, many skeptics took notice. Still, it was unclear how best to think about what was happening in the Arctic. Was this like the race to the moon—a matter of national pride requiring us to beat the Russians, or anyone else? Or was the Arctic a modern-day version of the Wild West?

In any event, there were enough old-fashioned border disputes to keep everyone busy. There was a border dispute between Canada and the United States over the legal status of the Northwest Passage; there was a dispute between Greenland and Denmark over economic and political independence; and there was a dispute among all Arctic nations over access to mineral rights. These are not things you bother to fight about unless you are fundamentally convinced that the ice will be gone someday. And yet, ironically, countries that had originally denied climate change were now scrambling over resources that would be valuable only if the climate models proved to be correct. This was proof that even sophisticated countries were still better at grasping short-term opportunities than responding rationally to long-term threats. There is nothing like a good old-fashioned gold rush to turn people into believers.

In 1946, the U.S. government was so impressed with the strategic potential of Greenland—the world's largest island—that it secretly attempted to buy this island from Denmark for $100 million. By 2011, the United States was wishing it had offered a lot more. The money to be made from aluminum production alone was enough to make your head spin. Alcoa had dammed up two rivers in West Greenland and built one of the world's largest aluminum smelters—mining about 340,000 tons a year. But Alcoa had to share its profits. Part of the deal was that the Home Rule Government of Greenland and Alcoa jointly owned the hydro power stations, the transmission lines, and the smelter plant. The people of Greenland knew a good opportunity when they saw it. And for them, climate change was a path to freedom. There was money to be made in the new Greenland. It was just a question of who was going to make it.

## May 2022

As the sea ice continued to melt, the disputes between Canada and the United States over control of the Northwest Passage increased. Canada maintained that the passage lay inside its territorial waters allowing it to exercise control over all ship traffic; the United States wanted the passage to be classified as an international sea-lane, outside any one nation's jurisdiction. Despite the controversy, the Canadian government staked its claim to the Arctic waters. It constructed new naval bases in the area and ordered a new fleet of Arctic patrol boats. It also established underwater listening posts for submarines and ships. In the end, Asia, Europe, Russia, and the United States refused to budge and the Northwest and Northeast passages were deemed international waters. Canada was, however, able to collect a small fee—about one-fifth of the $4 billion generated annually by the Suez Canal—for maintenance.

The retreating sea ice also opened a potential for deepwater drilling for oil and gas deposits. The Russians were holding tight to their claim that a large portion of the Arctic was their geological territory. But the gold rush philosophy of the "Wild North" could more accurately be described as "First come, first served."

As claims were staked and deals were made, fishermen a little farther south went in search of missing cod. The North Sea—the turbulent pocket of ocean bordered by the United Kingdom and Scandinavia (among others)—had always been a very fertile fishing ground. A little more than a century ago it yielded almost 20 percent of the world's fish harvest.

But by 2022, the North Sea was, on average, 3°F warmer than it had been fifty years before. The warmer temperature had chased away all the plankton that young cod eat in the spring. Just as the fishermen were searching for the cod, the cod were searching for the plankton. And the plankton were looking for cooler water.[26] In this new world, everybody was looking for something.

Because of climate change, nothing was where it was supposed to be, so international fishery management was a bit of a mess. In the meantime, the fish-and-chip shops in England were buying their cod from trawlers sailing the coasts of Iceland. The fishermen were trying to scrape by on shrimp and whelks (large marine snails). And the jellyfish had decided that the North Sea felt just right.

The telltale signs of a shifting climate were quite obvious in other ways, too. Across the lands of the Inuit, the formerly pale brown, treeless landscape had begun to turn dark green; spruce, larch, and fir trees were popping up from Sachs Harbor to Clyde River to Iqaluit. But the warmer temperatures meant serious trouble for any species that relied on the ice. Hunting had become increasingly difficult as the delicate food web continued to unravel. Seals and walrus depended on cod, the cod depended on crustaceans, the crustaceans depended on algae, and the algae depended on the ice to provide a home. In essence, the hunt for seals and walrus had become a hunt for ice. The amount of time hunters were able to spend out on the ice had shrunk from months to weeks to days as vast areas of open water made traditional hunting grounds inaccessible. Hunters were often forced to shoot some of their sled dogs because they had no walrus or seal meat to feed them. The "ice highway" the hunters had relied upon for centuries was closing down permanently.

Overall, the ground itself had become more and more unstable as permafrost thawed and gave way. Roads buckled, sewer and waterlines burst, home foundations sagged, and trees tipped over in random directions as if they were drunk. The permafrost was turning into a soggy sponge. There were countless attempts to find clever ways to keep the permafrost frozen, including refrigerated slabs and insulated carpets. But in the end chemistry always won. The heat was unstoppable.

## September 2032

The year 2040 was when climate scientists had collectively predicted the Arctic would be fully ice-free in summer. We all knew this was a conservative estimate. Most sea ice models, despite improvements in the physics and better satellite data, had repeatedly underestimated the speed of Arctic ice melt over the past decade. It was just a question of how much the models were underestimating the melt.

In the end, as natural climate variations such as the North Atlantic Oscillation and the human-induced long-term trend played tug-of-war, the models were off by only eight years. By 2032, summer sea ice in the Arctic had all but disappeared. As they patiently waited for the ice to recede, shipping companies had been carefully planning for the opening of the Northwest and Northeast passages. They reengineered the next generation of Arctic-ready cargo ships and trained crews in how to deal with the cold. When the time finally came, they were ready to take a test drive in the Northwest and Northeast passages.

The trip from Yokohama to New York via the Northwest Passage was 2,200 miles shorter than by the Suez Canal. And the Northeast Passage to Europe via the North Sea would save about 4,200 miles between Yokohama and Rotterdam. From Singapore to Rotterdam, it was about 1,300 miles shorter than going through the very expensive Suez Canal. But despite the focus on miles, the experts were quick to point out that it was *time* they really wanted to save. They estimated that the Northeast Passage could shave off about seven days of travel time.

But still, a lot of risks came with Arctic shipping routes. The worst was the need for absolutely open water so the ships would be able to maintain speed. If a ship hit even a small chunk of ice at 22 knots, the crew would have to deal with a hefty hole. Needless to say, there were still plenty of small chunks of ice floating around that were difficult to spot, even with the new remote sensing equipment. The first major oil spill occurred just a few weeks after the new routes opened, when an empty container ship heading back to Yokohama from New York collided with a chunk of ice moving in for winter. This was bad news all around; but it was especially bad for the fish, as they had fewer and fewer places to escape to.

The Norwegians had always billed their fish as the "purest fish in the world," thanks to the clean, cold Arctic waters. Uncontaminated fish were almost impossible to come by, and the Norwegians had been charging a premium. Needless to say, this wasn't a claim they could make anymore.

The warming had brought cod, herring, halibut, and haddock north in search of food, but it also brought oil and gas tankers and container ships from all over the world. In addition to rice from China and cars from Japan, the ships brought various contaminants and diseases that fouled the Arctic waters. Pollution and disease devastated the fishing stock. The Norwegian defense minister said in a speech at an international meeting in Moscow to discuss how to handle the collapse of the fisheries, "It used to be we had a world with plenty of food but bad distribution. Now we have a world where we have plenty of distribution options, but not enough food." That pretty much summed up the situation.

Meanwhile the Canadian government was moving full speed ahead with its plans to become the final resting ground for carbon dioxide. Carbon capture and storage facilities were built in British Columbia, Alberta, and Saskatchewan—which had been identified as offering the right combination of high-volume $CO_2$ emission facilities located close to abundant geologic storage sites.[27] The $CO_2$ would be liquefied and then injected into depleted oil and gas reservoirs, or into saline aquifers located more than 1 mile below the surface. The Canadians figured that they could sell space in their reservoirs and aquifers to the Americans as an additional source of income. Oddly, despite all the innovation during the past forty years, no one could figure out how to make $CO_2$ anything other than an expensive nuisance—the climatic equivalent of nuclear waste.

## October 2050

It's tempting to say that places like Greenland and Nunavut had been living their version of the American dream. The decision to move forward with the aluminum and zinc mines, the shipping, the oil tankers, and the natural gas pipelines turned Nuuk and Iqaluit—the capitals of Greenland and Nunavut, respectively—into fashionable international cities complete with four-star hotels and fine restaurants. But the boom had begun to show signs of busting—or perhaps more accurately, the boom was melting. Infrastructure costs to combat crumbling roads and sagging buildings were out of control. The permafrost was neither permanent nor frozen. And methane had begun to burst out at the Arctic's seams.

Methane hydrates, essentially natural gas trapped in ice crystals,[28] had become the next big thing in the Arctic, drawing investors from Dubai, Russia, the United States, and elsewhere. Imagine a snowball on fire and you've got a decent picture of what a methane hydrate looks like. Exploration for methane hydrates during the 2010s and 2020s had given way to large-scale production of natural gas during the 2030s. That was when the money really started to pour in. It was estimated that more energy was locked up in methane hydrates than in all other known fossil fuels combined. That was all most investors needed to hear.

The "Wild North" was now also dubbed "Saudi North." Hydrate deposits more than a half mile thick were found scattered under permafrost all over the Arctic, including the North Slope of Alaska, the Mackenzie River delta of Canada's Northwest Territories, and the Messoyakha gas field of western Siberia. Exploration also yielded several productive sites beneath the ocean floor at water depths greater than about 1,600 feet.

But as with everything else, the situation wasn't that simple. Methane brought tremendous wealth to the Arctic, but it also brought trouble. As temperatures continued to warm, methane hydrates, both in Arctic permafrost and beneath the oceans at continental margins, destabilized.[29] In other words, methane—a greenhouse gas with twenty-three times the heat-trapping capacity of $CO_2$, began pouring out of the Arctic. As the permafrost melted, the methane destabilization acted as a runaway feedback and further increased global warming. There was evidence that something like this had happened in the past, 635 million years ago, when unzippering the methane reservoir had warmed the Earth dramatically.[30]

Researchers had long warned that methane might be the final trigger setting off a climate change time bomb. New calculations showed that the levels of methane emissions from northern wetlands were going up year after year. And the potential for further warming was upward of several degrees—a scenario that frightened even the most stubborn skeptics.[31] In the end, many of us came to think that this new Arctic might after all have more in common with the barren lunar landscape.

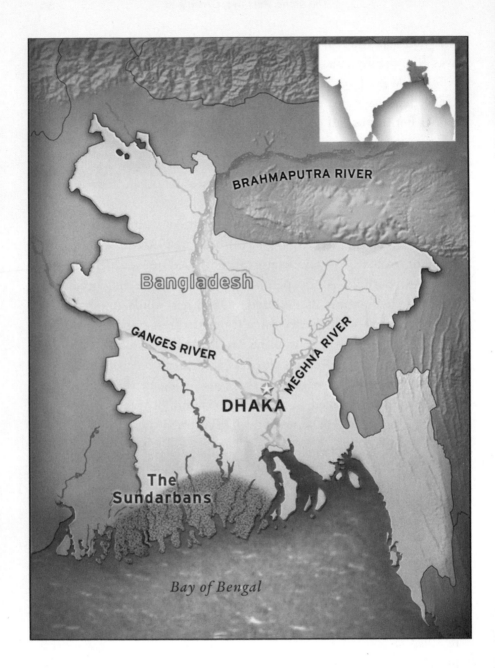

# 10

# DHAKA, BANGLADESH

It's hard to say why you fall in love with a person, or with a place. It's the same with science: sometimes your research is a lifelong passion, but sometimes a problem suddenly assaults you out of the blue and forces you to work on it—for the rest of your life. In the case of Peter Webster, an atmospheric scientist at the Georgia Institute of Technology, you might say his research on floods in Bangladesh began with a dare that led to a blind date that ended in a committed relationship. But it's probably better to just have him tell the story.

"It started off in a very strange way," he explains. "I was at a meeting in Bangkok in late 1998. We had just published a paper on the Indian Ocean Dipole." Webster's specialty is ocean-atmosphere interactions. The ocean and the atmosphere can interact in important and predictable ways that play a role in climate and weather: El Niño is a prime example of ocean-atmosphere interactions. Webster has a PhD from the Massachusetts Institute of Technology (MIT) and was among a crop of very successful graduate students who came out of MIT in the 1970s. Many of them, including Webster, worked with Jule Charney, a legend in numerical weather prediction. Webster is beginning to develop into a legend as well. He's received numerous scientific honors, including two of the most prestigious awards in the geosciences: the Carl Gustav Rossby Research Award from the American Meteorological Society (AMS) and the Adrian Gill Medal from the Royal Society. The wall of his office at the Georgia Institute of Technology in downtown Atlanta, where I met with him to talk about his research in Bangladesh, is a striking reminder of what the walls of accomplished people generally look like.

If you study climate science, you'll soon learn that there are a lot of *dipoles*: places where some quantity, such as atmospheric pressure or temperature, flips and flops between high and low or hot and cold. A dipole is a classic example of an ocean-atmosphere interaction. In the case of the Indian Ocean Dipole (IOD), ocean temperatures in the eastern and the western Indian Ocean flip between warm and cool, with the rains following the warm ocean temperatures.[1] The IOD also sets up what Webster calls "a seesaw of sea level" in the Bay of Bengal. Dipoles are important drivers of climate variability, and if you understand how they work, they can often help predict when a drought or a flood might be coming your way.

At the conference in Bangkok, Webster presented a paper suggesting a large-scale climate connection between the IOD, sea level in the Bay of Bengal, and the monsoon rains. This is exactly the kind of association climatologists search for, to better understand the physics of a system. Webster's research showed that when sea level in the Bay of Bengal was high, so was the risk of flooding: that is, the two events were highly correlated. "And in 1998," Webster explains, "the sea level in the Bay of Bengal was about a foot higher than normal."

At the time of his talk, Bangladesh was at the tail end of what came to be known as "the flood of the century." Sixty percent of the country had been flooded for more than three months, from July through September 1998. Dhaka, the capital of Bangladesh, was under 6 feet of water.[2] When all was said and done, the flood of 1998 caused 1,100 deaths; inundated nearly 39,000 square miles; made 30 million people homeless; damaged 500,000 homes; caused heavy losses to infrastructure;[3] and resulted in $2.8 billion in damages. Mother Nature has never cut Bangladesh a lot of slack.

Bangladesh has a problematic geography, and the floods are just the beginning. The geographic setting of Bangladesh makes the country highly vulnerable to many kinds of natural disasters. In addition to floods, it has experienced tropical cyclones, droughts, tornadoes, earthquakes, water contaminated by arsenic, and landslides. Bangladesh has seen more natural disasters than one might expect in such a small country—it is only about the size of the state of Iowa. But whereas Iowa has a population of 3 million, Bangladesh has a population of more than 162 million. That works out to some 2,900 people per square mile, making Bangladesh one of the most densely populated countries in the world.

Another problem is that two-thirds of Bangladesh is less than 17 feet above sea level; only in the extreme northwest will you find an elevation of more than 100 feet. And in this small, densely packed, low-lying country there are 230 rivers. Three of them—the Ganges, the Brahmaputra, and the Meghna rivers—come together to form a large floodplain. Eighty percent of Bangladesh sits within that floodplain, and everyone who lives there knows that in any given year, roughly one-quarter of the country will be flooded. And everyone also knows that every few years Bangladesh will experience a severe flood that inundates more than 70 percent of the country.[4] That's easy to predict. The hard part is predicting the details of where and when the floods will come. But one detail scientists are certain of is that climate change will make the floods worse.

And the floods already have the power to devastate. Farmers and fishermen can easily lose a year's worth of income during a single flood. These were the people Webster was trying to reach with his weather forecasts. "My research had suggested that the floods could be related to sea level in the Bay of Bengal. It was as if the floods came whenever the drain in the Bay got clogged up," Webster explains. So, at the end of his talk in Bangkok Webster made a rather provocative announcement: "We can now understand why Bangladesh has floods. And in understanding these large-scale controls, we can forecast them." Those were mighty big words—and Webster knew it.

The rains in Bangladesh begin in May, when the southwest trade winds, known as the *monsoons*, are drawn to the Indian subcontinent by the intense heat and consequent low pressure over Pakistan. The trade winds blow across the North Indian Ocean, picking up moisture along the way, before they head into Bangladesh and go through to the Himalayas. When the winds hit the side of the Himalayas, it begins to rain as a result of a process called *orographic uplift*. As the air travels up the side of the mountain, it cools, forcing the moisture to condense and fall out as rain. Basically, the rain continues until early October. During these months, the total rainfall varies from 4 feet in the northwest of Bangladesh to 11 feet in coastal areas, and to more than 16 feet in the northeast. Needless to say, this is one of the rainiest places in the world.

"And so I kept getting these e-mails saying 'Do you really believe you can predict the floods?' I said, 'Sure.' And I was very cocky about that," Webster says smiling. "Then they asked if I'd be interested in a grant to work on forecasting the floods. I said, 'Yes, of course.' So we wrote a proposal to develop a flood forecasting system for Bangladesh."

Professors, as a rule, don't turn down grant opportunities. And Webster saw this grant as a dare that had the potential to save hundreds or even thousands of lives each year. "I thought it would be a very easy problem, because all you really need for a flood forecast are four quantities: the forecasted rainfall, the sea level in the Bay of Bengal, and the levels of the two major rivers, the Ganges and the Brahmaputra, all the way to India," Webster explains. "In fact, I could never understand why the Bangladeshis hadn't already done it." So Webster and his team went off to collect the ingredients for their flood forecast model. This was his blind date with Bangladesh.

"I remember the first time I went into the villages," Webster says, thinking back. "I asked a farmer to tell me about the flooding. We were standing in his rice paddy, and he said, 'These fields here always flood because we are in the lowlands.'" Farmers living in these lowland areas have adapted to the floods by building their houses on raised mounds and adjusting the way they farm.

"Naturally, I asked him if he would take me to the highlands. At which point we walked a little ways over and up a bank of no more than 2 or 3 feet. He said, 'Here, these are the highlands.' " Webster says in amazement, "We were literally standing in a slightly raised paddy field. I would have never noticed the difference." But the farmer knew the difference very well. During a flood, it was the highland crop that might have a chance of surviving, saving him and his family from starvation.

When Webster and his team got to Dhaka to assemble the forecast model, they soon discovered that India provided no stream gauge data to Bangladesh for the rivers that originated up north in India. "That explained why no one had ever issued a forecast. No one had any idea what the conditions were upstream," Webster explains. Bangladesh may have a lot of rivers, but more than 90 percent of the water in those rivers comes from outside its borders. "The best we could do was a forecast out a day or two. And even that wasn't any good, because the upstream conditions are so critical," Webster says.

Data are the lifeblood of good research and the principal ingredient in a reliable forecast of weather or climate. That's why scientists work so hard at setting up international observing networks. The climate system knows nothing of national borders; and just as Lewis Fry Richardson discovered when he started the science of weather prediction on a battlefield during World War I, a model run with lousy data will give you a lousy forecast. Garbage in, garbage out.

"So we decided to build a hydrological model for both the Brahmaputra and the Ganges River basin that could *estimate* the streamflow upstream," Webster explains. They obtained data and weather forecasts from the European Centre for Medium-Range Weather Forecasts (ECMWF) and fed them into the hydrological models. Their model also incorporates estimates of precipitation from two satellite-based systems, along with streamflow measurements of rivers inside Bangladesh. "We basically used the rainfall forecasts from ECMWF to feed the rivers with rain. And then we used satellite data to calibrate the rainfall. That's how we worked around the problem. And we came up with a solid scheme for a ten-day forecast. Our first forecast was in 2003, and it was pretty darn good," Webster says.

In the summer of 2004, Webster and his team generated ten-day forecasts showing that the Brahmaputra River would be likely to flood on two occasions in July.[5] They were right. The 2004 floods inundated almost 40 percent of the country. But unfortunately their forecast had not been much help to the local people. At the time, the flood forecast still hadn't been fully integrated into Bangladeshi warning systems, and more than 500 people in Bangladesh and India died in the rising waters. So Webster and his team went about setting up a communication network. They worked with the Flood Forecasting Warning Center (FFWC) in Bangladesh and the Asian Disaster Preparedness Center to develop a network to distribute the forecasts directly to people in five districts along the Brahmaputra and Ganges rivers, including impoverished families living on islands that were known as river chars.

"It's all done using webs of cell phones, and it reaches over 100,000 people," Webster says. Cell phones are everywhere, even in remote villages. "People are told that a flood of a certain level is coming at certain time. And that allows farmers to harvest early and protect their seeds," he explains. Once the communication channels were improved, his forecasts began saving lives and livelihoods. "An economist we were working with calculated that people were saving, on average, 25 percent in infrastructure and household damage by knowing the flood was coming," Webster says. "And saving 25 percent of everything is a lot."

By 2008, after ten years of research and outreach, Webster and his team had proved they could make a skillful forecast. The next step was to operationalize the process. Webster wanted to assemble a team at the FFWC to produce and issue the forecasts. But the situation wasn't that simple. "I had a rather disturbing interaction with the director of the FFWC," Webster says. In May 2008, just as Bangladesh was entering the flood season, Webster set up a workshop in Dhaka where he taught meteorologists at the FFWC how to generate the ten-day flood forecasts.

"But the director told me the FFWC was hesitant to give out a ten-day forecast, even though we had demonstrated predictability," Webster continues. "We argued and he said, 'But what if the forecast is wrong?' " The director told Webster that the FFWC didn't want to be held responsible for any losses. "So I said to him, 'If you saw someone crossing a road just as a truck appeared over the horizon, would you wait to warn the poor soul until there was no chance for him to get away?' It was an unpleasant conversation."

Webster was shocked that someone would turn down an opportunity to give people enough lead time to prepare for and, it was hoped, prevent a disaster. "I think it's morally wrong. You can't do that." Webster remains committed to getting the forecast out to people in rural areas, people who need as much time as the science of forecasting can allow. "Right now, the FFWC has agreed to monitor the ten-day forecasts but will not issue them," he says.

It strikes me that maybe Webster's story isn't so different from the larger fight to communicate the risks of global warming. A forecast for 2050 is not so different from a ten-day forecast for flooding. Neither is perfect, but both are critical to making informed decisions. And many would argue that both carry the same moral obligation.

Omar Rahman is someone who has come to sense this moral obligation. Rahman is a demographer and physician who studies social networks and urbanization. "I'm not a climate change person," Rahman explains over the phone from his home in Dhaka. "I got interested in climate change from a different perspective. I am interested in migration." As it happens, Rahman is himself a living study in migration.

"I basically left Bangladesh to go to college in the 1970s and stayed in the United States for twenty-eight years," Rahman says. After studying biochemistry at Harvard, he received a medical degree from Northwestern University. In 1996, he returned to Harvard as an assistant professor of demography and epidemiology at the School of Public Health. His work focused on the concept of resilience—specifically, the resilience of people living in rural Bangladesh.

"I found it odd that I was sitting in Cambridge writing about rural Bangladesh," he says. Rahman had written extensively about issues of development and had begun to feel a pull back home. "I was an academic writing papers that very few people would read," he explains. He began to question the impact of his research. The pull of a place can be quite powerful, and Rahman began to contemplate returning to Bangladesh.

"A friend of mine gave me a very good piece of advice," Rahman says. "He told me, 'If you go back home, you need to go back for good. Don't just test the waters, because you'll never stay.' " Rahman explains, "Developing countries are difficult places to work, and constant comparison is detrimental to staying." And then came what might have been the last straw. "We were told we had to move out of our apartment in Cambridge," Rahman says, laughing. "And so, in some ways, it came down to either moving to the suburbs or moving back to Bangladesh." Rahman, his wife, and their two children moved back to Dhaka in 2003, the same year Webster issued his first flood forecast.

"Ultimately, I am interested in the people part of this whole thing," Rahman says. And it is the human part of the problem of climate change in Bangladesh that makes everyone very nervous.

The most widely used estimate of how many people around the world could become *climate refugees*, a term heavy with political and moral overtones, is 200 million by 2050. To put that in perspective, about 1 million Irish immigrants came to the United States because of the potato famine during the late 1840s. These projections about climate refugees are based on a very crude formula for estimating migration, so it's safe the say the numbers are still fuzzy. Modeling people's behavior is a lot harder than modeling a flood. But whatever the number ends up being, it's likely to be of a magnitude not seen before in human history. And in Bangladesh alone, the exodus is estimated to be in the millions. Projections range from 6 million to 15 million by 2050.

Most of the migration in Bangladesh right now is internal. People are moving from coastal and rural areas to cities such as Dhaka. The reasons for leaving the place where you were born are varied. However, "I will honestly tell you that right now, most of the migration is economic," Rahman says. "None of the migration is driven by concern about climate change. That will come in twenty to thirty years."

But some observers argue that climate migration is happening already, and that it's being blamed on the weather instead of the climate. More severe floods and droughts are hitting the landless and poor farmers in the same villages that Webster visited. And the floods and droughts, which have always occurred in Bangladesh, will grow worse with time. Increased soil erosion and saltwater intrusion in coastal areas will make it more difficult to farm and work the nets for shrimp fry, leaving people with few options but to migrate. Many will end up in the slums of Dhaka. In the end, of course, it's always the economy. In rural Bangladesh, the weather is the economy. And if you believe the climate models, the weather will get worse.

By 2050, the population of Bangladesh will have grown from about 162 million people today to more than 220 million.[6] Today, more than 13 million people live in Dhaka. It's the fastest-growing megacity in the world. And every year, slightly more than 400,000 people in Bangladesh move to the capital, hoping to find a better life. Nearly 15,000 new cars were sold in Dhaka in 2008, a record high. There may be plenty of people and cars, but there are acute shortages of just about everything else. There are no sidewalks.

There is no mass transit system. And right now, there is enough power for only about 35 percent of the population. When I spoke with Omar Rahman on the telephone, he had been without power for eight hours that day. As he said, comparison with the developed world is detrimental. But nonetheless, when people along the coast who are unable to grow rice or work the nets to catch shrimp fry make the comparison between Dhaka and their own situation, they will still decide that Dhaka holds the keys to a better life. And by 2050, this megacity with very little energy, transportation, and water infrastructure is expected to be the home of more than 40 million people.[7]

Experts like Rahman worry about how Dhaka will cope with the rapid and unplanned urbanization in Bangladesh. Dhaka is not immune from the problems of geography that plague the rest of Bangladesh. The city is located in the coastal zone and is just as vulnerable as the rest of the zone to floods, storms, and tropical cyclones. Drainage is already a serious problem, and sewers routinely overflow during the monsoon season. And the slums, situated in the lowest-lying parts of the city, are even more vulnerable. The millions of poor who have settled there are crowded into metal shacks with no running water. They have merely traded one form of vulnerability for another.

"The impact of rapid urbanization is huge," Rahman says. "We are trying to model it because we know there is no way to stop it. By 2050, Bangladesh will have transformed from a population where only 5 percent of people lived in urban areas in the 1970s to a population where 50 percent of people live in urban areas." Rahman explains that the growth in Dhaka is a microcosm of the urbanization taking place across Africa and Asia. The UN Development Program estimates that by 2050, 70 percent of the world's population will be living in urban areas.

The next step in the migration pattern is across national borders. "When most people think of these issues," says Webster. "They mostly think about how terrible it's going to be for the people there. But this is an enormous problem from a national security standpoint, too. Because all of a sudden you have 200 million people who are displaced, people who have become climate refugees. Where do they go? They go to India. They go to Myanmar. But they won't be very happy people," Webster concludes.

For national security experts, migration is one aspect of climate change that evokes a real sense of dread. In the global landscape the argument is simple: what happens in Bangladesh doesn't stay in Bangladesh. The United Nations Department of Economic and Social Affairs estimates that 1.1 million migrants will enter the United States each year between now and 2050. Many are expected to come from Bangladesh.

It's also estimated that more than 10 million Bangladeshis have already made the move to India during the past twenty years. This issue is a constant source of tension between the two nations, and climate change isn't helping. India, for its part, sees climate change as bringing multiple threats, and apparently one threat is people. And so India is in the process of building a fence to keep them out. "I think they were going to build a fence anyway," says Rahman. "The data on this isn't clear, but I think the fence was ultimately built for political reasons. And the climate refugee argument is being used as an excuse," Rahman adds.

India maintains that the purpose of the fence is to protect the country against smuggling and terrorism as well as illegal immigration, claiming that about 5 million Bangladeshis are in India illegally. This is a number the government of Bangladesh is quick to contest. The fence runs along India's porous 2,500-mile border with Bangladesh. It is high, and it's made of heavily reinforced barbed wire. Climate change may not have created the fence but provides a plausible reason to continue building it. Still, as the Indian government works to complete the fence in the hope of keeping people out, the problem is not so much the people as the climate. The fence won't stop the floods, the cyclones, the droughts, or the rising sea level. The forecast for 2050 is going to require a lot more than a barbed wire fence.

————

Water, now and always, is at the heart of Bangladesh's problems. The country suffers from both too much and too little water.

The scenarios for 2050 and beyond predict that this water problem will worsen, owing to a number of factors. Rising temperatures and decreasing winter precipitation will bring more drought. Rising sea level will bring salt water into the rice paddies and rob Bangladesh of its agricultural land. Floods that result from increased snowmelt and a stronger monsoon will happen more frequently and last longer. And more intense cyclones will batter the coast, an economic hub and cultural treasure, with higher storm surge. This small country has always been vulnerable, and climate change will make it more vulnerable. It doesn't take a climate scientist to figure that out. Just ask Omar Rahman.

"I came into this field as somewhat of a climate skeptic," Rahman admits. "As a scientist, I always want to see the data. But I have to say I was convinced. These were serious people, people who were not prone to exaggeration. And the data spoke for itself," Rahman says. The Intergovernmental Panel on Climate Change (IPCC), the scientific group responsible for building the data and models that convinced Rahman, has issued a very strong statement about the changes that are taking place in Bangladesh. Temperatures in Bangladesh have already increased. The Fourth Assessment report indicates an increasing trend of about 1.8°F in May and 0.9°F in November during the fourteen-year period from 1985 to 1998. Annual average temperature in South Asia (5°N to 30°N, 65°E to 100°E) is projected to increase 3.2°F by 2050 and 5.6°F by 2100, according to the IPCC Fourth Assessment Report. The seasonal values for South Asia are shown in the accompanying table. Temperatures in Bangladesh are projected to increase 1.8°F and

2.5°F by 2030 and 2050, respectively, according to a recent assessment by the Bangladesh Centre for Advanced Studies (BCAS).

|  | 2020s | 2050s | 2080s |
|---|---|---|---|
| **Dec.–Feb.** | 2.2°F | 5.8°F | 9.7°F |
| **Mar.–May** | 2.2°F | 5.4°F | 9.4°F |
| **Jun.–Aug.** | 0.9°F | 3.0°F | 5.6°F |
| **Sept.–Nov.** | 1.3°F | 4.3°F | 7.6°F |

Projected changes in surface air temperature for South Asia under the highest future emission trajectory (A1F1) for three time slices: 2020s, 2050s, and 2080s. (SOURCE: IPCC FOURTH ASSESSMENT REPORT)

These warmer temperatures spell trouble for the Himalayan glaciers, the largest body of ice outside the polar caps and a critical source of freshwater throughout Asia. About 15,000 Himalayan glaciers support perennial rivers including the Ganges and the Brahmaputra that, in turn, provide a lifeline to millions of people in Bangladesh and across Asia. Himalayan glaciers are receding faster than glaciers in any other part of the world; and if the planet keeps warming at the present rate, the likelihood that they will be disappearing by the year 2035, and perhaps sooner, is very high.

The roughly 20-mile-long Gangotri glacier has been receding alarmingly in recent years. Between 1842 and 1935, it was receding at an average rate of 24 feet every year; the average rate of recession between 1985 and 2001 was about 75 feet per year. The current trends of glacial melt suggest that the Ganges and Brahmaputra, which crisscross the northern Indian plain, could run dry during the summer months in the near future as a consequence of climate change. And the IPCC report indicates that India will reach a condition of water stress before 2025. A fence won't be of much help when that happens.

Whereas Rahman may have initially been a skeptic, the people who live in rural Bangladesh have watched the climate change with their own eyes. According to a study carried out by the International Union for Conservation of Nature (IUCN) in Bangladesh, people in rural communities are reporting excessive and erratic rainfall, an increase in the number of flash floods, temperature variation, changes in seasonal cycles, and an increased occurrence of droughts and dry spells. These effects are likely to worsen, and adaptation strategies are urgently required. The question is: how bad does a situation need to get before it makes people leave?

One complication is the fact that, like floods, droughts in Bangladesh are seasonal. Depending on their timing, they can devastate crops, especially in the northwestern region, which generally has lower rainfall than the rest of the country. Drought brings significant hardship to poor agricultural laborers and others who cannot find work. In these areas, unemployment leading to seasonal hunger is often a problem, especially in the months leading up to the November–December rice harvest. If the entire crop fails because of drought, the situation for poor people can become critical. The IPCC predicts lower and more erratic rainfall, resulting in increasing droughts, especially in the drier northern and western regions of the country.

Too much water is no better. The problem is, again, the timing. More than 80 percent of the roughly 7 feet of annual precipitation in Bangladesh comes during the monsoon period. Most of the climate models estimate that precipitation will increase during the summer monsoon. The reason is fairly straightforward. As the temperature increases, air over land will warm more than air over oceans in the summer. This will deepen the low-pressure system over land, which happens anyway in the summer, and will enhance the monsoon. Moreover, there is a link between Eurasian snow cover and the strength of the monsoon; when snow cover retreats—as it is expected to do with the higher temperatures—the monsoon strengthens.

Climate models indicate a general increase in the intensity of heavy rainfall events in the future, with large increases over the Arabian Sea, the tropical Indian Ocean, northern Pakistan, northwest India, northeast India, Bangladesh, and Myanmar.[8] Scientists expect a roughly 10 percent increase in rainfall during monsoon the season by 2030, while dry seasons could see harsher droughts.

The problem for Bangladesh is that two things are happening at once. The warmer temperatures are intensifying the monsoon and they are melting the glaciers; and unfortunately the melting season happens to coincide with the monsoon season. Rapid glacier melt will mean more water flowing down the Ganges and Brahmaputra rivers during the monsoon months, causing even more devastating floods. Webster has been studying the changing frequency of flood events. "We've been looking at the flow of the Brahmaputra and the Ganges out to the year 2100. Two things happen. You see more flood events and the flood levels are much higher," Webster explains. Scientists say once-in-twenty-year floods are already occurring about every four years. However, in the long term, as the water in the rivers disappears, the result will be more severe droughts.

Although floods and droughts are a serious concern, the most serious issue of all is a rising sea level. The 146 million people living within about 3 feet of mean sea level worldwide are at risk from the projected rise in sea level over the coming century. An even greater number—268 million living within about 16 feet—are at risk when the added impact of storm surge is considered. Moreover, these numbers are rising, owing to the combination of a growing population and its coastward migration. A 3.3-foot rise in sea level would inundate about 20 percent of Bangladesh's total land, directly threatening 11 percent of the population with inundation (this figure is based on current population distribution). In addition, the backwater and increased river flow from sea level rise could affect 60 percent of the population.[9]

Oddly, the rise in sea level is the one estimate for which the cautious ways of the IPCC that Rahman so appreciated are wrong. In an effort to address the uncertainty, the IPCC chose to go with the most conservative estimate, which some would say is wildly conservative. The IPCC Fourth Assessment Report estimated global sea level rise over the coming century in a range from 11 inches to 31 inches. But it left room by stating, "The upper values of the ranges given are not to be considered upper bounds . . . for global sea level rise because existing models are unable to account for uncertainties such as changes in ice sheet flow."

Even with these conservative estimates, the IPCC Fourth Assessment Report still projects that rising sea levels could wipe out more cultivated land in Bangladesh than anywhere else in the world. In Bangladesh, you can drive 60 miles inland from the coast, and you'll go up only a few feet in elevation. When you get about 150 miles away from the shore, the land finally starts to rise up another 50 to 65 feet. I recently talked with experts who are preparing the next IPCC report; they said the projected value rise in sea level is actually closer to 5 feet, instead of the previously estimated 31 inches. It's tough to wrap your mind around that. The factor driving migration is the land loss in coastal areas that will result from the rising sea level. "Somewhere between 20 to 25 percent of Bangladesh will be inundated in the next fifty years," Rahman says. In Bangladesh, even the land is leaving.

The main impact of a rising sea level would be salinity ingress, causing the rivers in the coastal belt to become brackish or saline. This would have a serious impact on food production. In Bangladesh, production of rice and wheat might drop by 8 percent and 32 percent, respectively, by the year 2050.

Rising salinity levels as brackish water inundates cropland could hurt rice and wheat production. Overexploitation of groundwater in many countries of Asia has resulted in a drop in its level, leading to an ingress of seawater in coastal areas and making the subsurface water saline. India, China, and Bangladesh are especially susceptible to increasing salinity of their groundwater as well as surface water resources, especially along the coast, due to increases in sea level as a direct impact of global warming. The Meteorological Research Center at the South Asian Association for Regional Cooperation carried out a study on the recent rise in coastal sea level in Bangladesh. The study used twenty-two years of tidal data from three coastal stations. It revealed that the rate of rise in sea level during the last twenty-two years is many times higher than the mean rate of global rise over 100 years; this suggests that regional subsidence could be making the situation worse.

There are also the tropical cyclones. According to the UN Development Program, Bangladesh is the most vulnerable country in the world to cyclones. Scientists report that 2007 was the worst year on record for intense hurricanes in Bangladesh. "The worst-case scenario for Bangladesh is rising sea levels and increased floods," says Webster. "But you add to that, of course, increased intensity of hurricanes in the spring and fall. So that would be the triple whammy.

"I'm not quite sure if Bangladesh is an adaptable country," Webster continues. "Imagine a country the size of Iowa becoming half the size of Iowa with double the population." That is the long-term forecast for Bangladesh. "Ultimately," Webster explains, "they are running out of land."

And some of the land is so beautiful. The Sundarbans—the wetlands region straddling the coasts of western Bangladesh and neighboring India—was formed by the deposition of materials from the Ganges, Brahmaputra, and Meghna rivers. If the Sundarbans is lost, the habitat for several valuable species will also be lost. An 18-inch rise in sea level would inundate 75 percent of the Sundarbans; a 26-inch rise could inundate all of the system. That would threaten what is now the single largest mangrove area in the world and is designated a World Heritage Site. The name *Sundarbans* means *the beautiful forest* in Bengali. The mangroves in this forest, within the delta of the Ganges, Brahmaputra, and Meghna rivers on the Bay of Bengal, act as natural buffers against tropical cyclones and as a filtration system for estuarine water and freshwater. They also serve as nurseries for many marine inverte-brate species and fish. The Sundarbans mangrove forests are well known for their biodiversity, including 260 bird species, Indian otters, spotted deer, wild boars, fiddler crabs, mud crabs, three marine lizard species, five marine turtle species, and several threat-ened species, such as the estuarine crocodile, the Indian python and the famous Bengal tiger. It was for these reasons that the Sun-darbans National Park, India, and the Bangladesh part of the Sun-darbans were added to the World Heritage List in 1987 and 1997, respectively.[10]

The rise in sea level and the decreased availability of freshwater—particularly during winter, when rainfall will be less—will cause an inland intrusion of saline water. As a result, many mangrove species, intolerant of increased salinity, may be threatened. In addi-tion, the highly dense human settlements just outside the mangrove area will restrict the migration of the mangrove areas to less saline land. The mangroves are caught in the middle. The shrinking of the mangrove areas will have an effect on the country's economy. Many industries that depend on raw materials from the Sundarbans will be threatened with closure, and large-scale unemployment could result. A project of the United Nations Development Program (UNDP) has evaluated the cost of building about 1,375 miles of protective storm and flood embankments that would supposedly provide the same level of protection as the Sundarbans mangroves. The capital investment was estimated at about $294 million and the yearly maintenance budget at $6 million—much more than the amount currently spent on the conservation of the mangrove forests in the area.

Webster may be a skeptic who thinks Bangladesh may not be adaptable, but Omar Rahman has no choice but to find adaptation strategies that will work for this country. It is his home. When I ask him what the year 2050 might look like if we stick with the status quo he is quick to respond. "I think if climate change is not taken seriously, if the predictions are right and sea level rises more than one meter, we would see an almost unimaginable catastrophe. It's the worst-case scenario," says Rahman.

He thinks the rapid urbanization exacerbates the problem. When 50 percent of a country is urban, it is no longer a rural country. And Rahman agrees that food production is problematic. "Right now the country is almost self-sufficient in rice production—which is pretty amazing, given that a country the size of Iowa can feed over 160 million people. But we will lose a lot of acreage, and that means we'll have to import a lot of rice, like Japan. From an economist's point of view, that's going to be a big shock, but not something that is unapproachable, if we start thinking about it now. Better than if we sit and wait.

"It's unfortunate that we seem to be beset by natural disasters. But when I think about the story of Bangladesh, I think it is a story of hope," Rahman says.

Bangladesh arose in 1971 from a civil war. At the time, the U.S. secretary of state, Henry Kissinger, called Bangladesh an "international basket case."

"Now I like to say Bangladesh has gone from being known as an international basket case to a Bengal tiger," Rahman says. Until the worldwide economic slump that began in 2008, Bangladesh's economy was growing at a pace not far behind India's; Rahman attributed this to a developing culture of entrepreneurship and a thriving garment industry. Garments are the main export for Bangladesh, making up more than half of total exports.

"I don't think anyone would have predicted that we would have become an entrepreneurial culture. The garment industry has bred a new class of entrepreneurs," Rahman explains. In 2007, the World Bank predicted that Bangladesh could join the ranks of middle-income countries within two decades. "The best thing that's happened to us is that we don't have oil or mineral wealth," Rahman says. "We've had to develop our people. And we've done spectacularly well, despite our natural disasters—famine, you name it, we've had it."

The economic growth has happened because of the resilience and the tenacity of the Bangladeshi people. But Rahman would still like to see Bangladesh become more sustainable; that's one of the reasons he came back.

"As a demographer I'm interested in structure. It has had the most successful family program. In the mid-1970s women had seven children; now it's about 2.6." Estimates project that the population should stabilize at around 250 million. That still is a lot people to feed on a plot of land that is only the size of Iowa, and which is expected to shrink by one-fourth, owing to a rising sea level, by the middle of the century.

Rahman has an idea that he thinks might help kick-start adaptation programs: an international center for climate change adaptation that's actually located in the developing world. As he says, "You know you are underdeveloped when most of the literature about your country is written by people outside your country. We decided to set up the center *inside* the developing world."

Rahman believes students will learn more in the living laboratory of Bangladesh than in a sterile classroom in Cambridge about what vulnerable countries need to tackle climate change. Rather than getting hung up on the fence, the country needs to build embankments. It needs cyclone shelters and research on rice. And it needs to address the already explosive internal migration to Dhaka.

"I said: Look. We as a university would like to sponsor a coordinated set of activities. Train the next generation of scientists. Not enough people have the right kind of skills. It will be international. Draw people from all over the world as well as from inside Bangladesh. We will train people to think on timescales climate change requires," Rahman says. The program at the Independent University, in Dhaka, where Rahman is provost, will be a twelve- to eighteen-month master's program. Students will spend time out in the field, as Webster did. And their research will cover all aspects of the problem. Climate change is a problem that will require engineers and computer scientists as well as chemists and anthropologists to solve. Climate change is not a problem that can be constrained in a straitjacket.

"As an educator, the best part is that you get a chance to mold young people. For decades now, people have gone to America because it offered so many opportunities. It's changing now. There is an insularity. America has always struggled with being open versus retreating into itself. I benefited so much so I don't want to see that change. The thing that I miss the most is the ability to reinvent yourself. That's the best part of American culture. America is the most egalitarian place in the world. It's the sense that you can be anyone. You don't feel that you're worse off. I want Bangladesh to move to that."

Rahman seems to be suggesting that if he can help his students reinvent Bangladesh, they may in turn reinvent themselves. This is what happens when a trained psychiatrist works on global warming.

As Rahman looks to the future, he sees three adaptation options for Bangladesh: retreat, accommodation, and protection. "In view of the high population density and shortage of land, retreat is not possible. We should pursue the two other options. Some of the adaptation options are: raising of forest all along the coast, protection of mangrove forests, changing cropping pattern and variety in the coastal area, construction of embankments where feasible, construction of 'safe shelters' for emergency situations like extreme events, etc. In fact, many of these options are already in operation—on a limited scale, though.

"Adaptation has crosscuttings of different disciplines, and hence a multidisciplinary and integrated approach need to be taken up to reduce vulnerability. Adaptation will require thinking big and small—for instance, changing cropping patterns and developing new seeds able to survive in the changed climatic conditions. Part of our attempt is to train leaders to prepare before it happens. . . . You need to mobilize people and you need to empower them," Rahman says.

In the end, it's the people who have the potential to do this. They

will improve the adaptive capacity of the country. Bangladesh may
have a problematic geography, but it has great people—people who
refuse to let the West define them as hopeless.

Leaders here estimate that it will cost $500 million just to raise
embankments in some areas about 8 inches, a level that by the time
construction is complete might not be high enough to keep growing storm surges at bay. Adaptation is not sexy or cheap. Scientists
at the Bangladesh Rice Research Institute are working to develop a
strain of rice that can withstand higher salinity levels.[11] Adaptation
will require infrastructure investments across the board. Bangladesh
needs to build embankments and cyclone shelters. The government
says the country's power-generating capacity is at a maximum,
4,000 megawatts, which covers only 35 percent of the total population. The newly elected government has vowed to increase power
generation in order to boost economic development. Rahman estimates that by 2015, the demand for electricity in Dhaka will rise
to 10,000 megawatts. That, Rahman says, creates an enormous opportunity for clean energy projects to promote energy efficiency and
renewables at a household level.

But he noted that resistance to spending precious dollars on more
expensive low-carbon technologies in Bangladesh remains strong.
Here, economic growth and fighting poverty remain the top priorities. "We are one of the most negligible emitters of greenhouse gas,"
Rahman says. Bangladesh, currently one of the poorest countries on
Earth, has virtually no hand in causing climate change. The average Bangladeshi emits about one-third ton of carbon dioxide each
year—a lot less than the roughly 20 tons emitted annually by the
average American. At the global level, Bangladesh emits less than
0.2 percent of world total. To put that in perspective, the city of
New York alone emits about 0.25 percent of the world's total greenhouse gases.

As Rahman says, "Cooking stoves account for almost 20 percent
of emissions in Bangladesh. Cooking stoves. This is the level of
industrialization we're talking about." Rahman still has hopes for

megacities; he says that leaders need to start viewing land use and other aspects of city planning as critical components of preparing for climate change. "Properly managed, urbanization can be a good thing," he said. "Improving urban management is itself an adaptation strategy."

For people in Bangladesh, climate change is not a theoretical, academic, or distant concern. It is a question of survival. It is a question of infrastructure. It is a question of water and energy. It is a question of believing the forecast for 2050 and beyond.

At the end of our long phone conversation, Rahman couldn't help making a comparison between his old life in the United States and his new life in Bangladesh, "I lived in the United States for twenty-eight years, and there are things I miss. There is no question about that. I would have had more things if I stayed in the United States: a bigger house, a bigger car. But even that . . . I don't regret for an instant. I think I've come at a very exciting time. But there is no question in my mind that I made the right decision. This is not just about me; it's about something larger than myself. I don't think I could say that if I had stayed in the United States."

I guess sometimes you do get to choose whom you fall in love with.

## Bangladesh: The Forty-Year Forecast—Sea-Level Rise, Floods, and Climate Refugees

| DHAKA, BANGLADESH | TODAY | | 2050 | | 2090 | |
|---|---|---|---|---|---|---|
| Emissions Scenario | JAN. | JULY | JAN. | JULY | JAN. | JULY |
| Higher | 65.6 | 83.7 | 68.4 | 86.2 | 72.9 | 88.8 |
| Lower | | | 68.3 | 85.6 | 69.7 | 86.6 |

## Forecast
## January 2016

Nowhere was the issue of water more problematic than in South Asia. From the 29,029-foot Mount Everest in the Himalayas down to the lush, swampy mangrove forests of the Sundarbans, water was being held hostage by climate change. And that meant more than 1.3 billion people, dependent on the good graces of the climate system to deliver life-sustaining water in a timely and dependable manner, had a very big problem on their hands.

The problem came in several forms and started at the very top, in the Himalayas—mountains that stretch from Pakistan to India, China, Nepal, and Bhutan. The more than 15,000 glaciers that have covered the Himalayas for millennia bear a significant responsibility—they feed Asia's nine largest rivers, including the Ganges, Indus, Brahmaputra, Mekong, Yangtze, and Yellow, and bring a steady supply of pure, cool water to the people of South Asia. The problem was that the glaciers draped over these majestic mountains were retreating at an alarming rate. Scientists estimated that most were pulling back between tens and hundreds of feet each year; this rate made the Himalayan glaciers the fastest-melting glaciers in the world. In this very remote and beautiful place, the sound of global warming had become deafening.

The sound itself came from what is known as a glacial lake outburst flood (GLOF). Glacial lakes, which form as a result of a melting glacier, had become overwhelmed by meltwater. Every few minutes the chemistry of global warming showed off its handiwork: somewhere along the 1,500-mile mountain chain, rising temperatures ripped heavy chunks of ice loose from glacier after glacier. As the ice came loose, it crashed down, adding more and more water to already overflowing glacial lakes. Eventually, the lakes had no choice but to burst—releasing huge quantities of water. By 2016, every country in the Himalayan region had suffered from glacial lake outburst flooding.[12]

The melt rate had begun to increase in the early 1990s. It was then that the now infamous Luggye glacier in Bhutan—retreating more than 520 feet a year—finally broke off, on October 7, 1994. The lake burst open, releasing more than 4 billion gallons of water down the Pho River, killing 21 people, and wiping out entire villages and farms.[13] Floods like that were almost routine by 2016.

After countless floods, local villagers in Bhutan and across the Himalayas took matters into their own hands. They organized a small army of workers to combat the effects of climate change. Local officials estimated that by reducing glacial lake levels by 15 to 20 feet, they might be able prevent catastrophic flooding. The cost of widening just one lake ran upward of several million dollars. For the most part, the money was provided by the Least Developed Nations Fund—a special fund set up by the United Nations Framework Convention on Climate Change to help the world's poorest nations adapt to climate change. In Bhutan the work was done without the help of heavy machinery—the workers used just picks and shovels. In their first year, people from across Bhutan working at the Thorthormi glacier managed to lower the lake's level by 35 inches. But it would take years to get the lake to a safe level. And while there was a great sense of accomplishment with each inch the lake dropped, people realized that the risk of flooding would eventually be dwarfed by the problem standing in line behind it. The bigger catastrophe would come with the eventual disappearance of the glaciers. After the floods came the drought.

## April 2022

Problems with water came from the mountains, but they also came from under the Earth. Groundwater—formed by the natural percolation of rain and snow into soil and stored in pockets of porous rock—was being depleted at an alarming rate. Satellites operated by NASA first revealed that groundwater levels in northern India had been declining by as much as 1 foot per year over the past two decades—a completely unsustainable rate. The Indians denied this, but the satellites did not understand politics. They clearly showed that 26 cubic miles of groundwater had disappeared from aquifers in areas of Haryana, Punjab, Rajasthan, and the nation's capital territory, Delhi, over the past four years alone—enough water to fill Lake Mead, the largest man-made reservoir in the United States, three times over.[14] The Indian government said that U.S. satellites should mind their own business.

The reductions of streamflow and groundwater did little to improve relations between India and Pakistan. A meeting was called between high-level Indian and Pakistani officials to renegotiate the Indus Waters Treaty, first signed by the two countries in 1960. Upon signing the treaty, the countries had agreed that the six primary rivers of the Indus basin would be split evenly between India and Pakistan. For more than sixty years the treaty withstood the strain of wars between India and Pakistan. But now that groundwater had also become scarcer and scarcer, Pakistan believed it was being taken advantage of and requested a larger share of Indus water. The fact that both countries had nuclear capabilities lent a chilling new dimension to the water negotiations.

## August 2026

Woes involving water in the south came in a different form. In early August, millions of Bangladeshis had been marooned or displaced by floodwaters. The death toll currently stood at more than 5,000, but it was expected to rise. The floods had been coming more often, just as the models had predicted. Everyone was grateful for the seasonal and twenty-day flood forecasts routinely issued to Bangladeshis by the Asian Disaster Preparedness Center. At least these forecasts gave people a chance to prepare. With advance notice, they could postpone planting or hurry to harvest some or all of their crops, move livestock to safety, encircle fishponds with nets to prevent fish from escaping, and stock food and other supplies. It was something. And it allowed them to continue living in the place they loved.

In addition to relying on the flood forecasts, villagers also engaged in simple solutions they hoped would help decrease their vulnerability. Home foundations and frames were constructed using lightweight composite materials that could bend but would not break during a storm. Women wove these fibers from jute, one of Bangladesh's common plants, with recycled plastics to form strong building material. The people of Bangladesh did everything in their power to stay. They even used materials that would float on the rising tide of a coastal surge—hoping these might serve as life rafts when the next flood came. But in the end, it was not their choice to make. With each successive flood more and more people began to pack what was left of their possessions and leave. First the numbers were in the thousands; then they gradually increased to the hundreds of thousands. A steady rain of refugees poured down on Dhaka—a city already overcrowded. Men and women sometimes arrived with dreams of finding better jobs and better lives. But many had left their villages with no dream other than to save the lives of their children—too many had been lost to the floodwaters already.

## September 2039

In India, the groundwater situation worsened. Officials were forced to begin illegally withdrawing more water from the Indus River than they had been allotted. In response, Pakistan threatened to call in troops along the border.

As Himalayan glaciers retreated and groundwater was further drawn down, sea level continued its steady rise.[15] As the sea level crept higher and higher, the saltwater front traveled hundreds of miles upstream, and the salinity in surface water increased almost sixtyfold. The increase in salinity altered soil quality and nutrient loads. Simply put, the salt water was killing the trees. Down in the Sundarbans, the mangroves were dying. And the ripple effect of climate change was not hard to predict. The mangrove forests of the Sundarbans contained one of the last remaining populations of wild Bengal tigers left in the world.

## July 2050

Despite plenty of competition, South Asia remained the most food-insecure place on planet. Rice and wheat yields continued to spiral downward because of high temperatures and low water supply.[16] Governments around the region had tried to pool resources and engage in some of the less expensive adaptation measures. The hope was to moderate the predicted crop shortfalls and keep as many people fed as possible. As a result, planting dates were shifted and farmers switched to existing drought- and heat-tolerant crop varieties. Money was also spent on the development of new crop varieties and the expansion of irrigated areas. These measures showed some the biggest benefits. But in the end, there is only so much you can do without water. With an additional 130 million people pushed from food insecurity to famine, a mass exodus was under way. As part of an international agreement, the United States and the European Union agreed to take in millions of the hungry and displaced. And after years of fighting over what to call them, they were now officially known as climate refugees.

In Bangladesh, the water problem was surreal. With almost 25 percent of the country underwater—as a result of rising seas, recurrent flooding from increasingly malicious tropical cyclones, and the slow and deadly seepage of saline water into wells and fields—Bangladesh had become a wasteland. One by one, millions of men gathered their families and left their mud-caked villages, never to return. Many of them crossed illegally into India, hoping to find construction work in Assam and West Bengal.

Others, like Hassan and his family, fled to Dhaka. In his village Hassan had been a proud man, able to support his family with his handiwork. Now he was shining shoes on the streets of this megacity, whose population had swelled to more than 40 million. Hassan and his family were among the last to leave their village. It took him several months to persuade his wife to move away from the only place she had ever known. But finally, she gave in. He promised to take her to the Dhaka Zoo. She had always dreamed of seeing the Bengal tigers up close. The zoo was now the only place they could be found.

# 11

# NEW YORK, NEW YORK

New Year's 2000 is one of the few moments in my life that I remember with great precision. I was standing on the corner of Fifty-Ninth Street and Columbus Avenue, freezing cold but happy to have company. My roommate had decided, uncharacteristically, that she was going to Times Square with some friends, and so I tagged along not wanting to be home alone in the apartment. It was almost midnight, so Fifty-Ninth Street was as close as we could get to Times Square. Despite frigid temperatures and months of media hype with dire predictions of power outages, bank runs, and other random catastrophes, there we all were. Hundreds of thousands of people from all over the world, ready to hug a stranger and ring in the New Year. As always, there was a sense that the New Year—in this case, it was also celebrated as the start of a new millennium—held the possibility of something better. And as far as I could tell, the competing sense that we might also witness human civilization crushed by a little bug was not lost on anyone.

That bug, of course, was the Y2K bug, the millenium bug. It was nothing more and nothing less than a computer bug resulting from the practice in early computer program design of representing the year with two digits. In fact, the term *Y2K* itself was born out of a good programmer's relentless pursuit of efficiency. David Eddy, one of an army of programmers who worked on fixing the problem is credited with coining this term. He says he coined it on June 12, 1995,[1] in an absentminded e-mail. "Being a good programmer," he explains, "I'm a minimalist typist. And Y2K was simply 60 percent less effort/cheaper to type than year 2000." Funny the way good intentions can come back to haunt us.

Actually, Y2K was wrapped up in something much bigger. "Y2K coincided with the end of the millennium, so it became somewhat of a Rorschach blot for our collective anxiety about the future. The greater the number of 0s in a year, the more we freak out," explains Paul Saffo, a technology forecaster based in Silicon Valley who was among the first to push businesses to take Y2K seriously. He adds, "Y2K tapped into some pretty apocalyptic stuff. And in that sense I think it has some similarities with climate change." Saffo is a consulting professor in the School of Engineering at Stanford University, where he teaches forecasting and the impact of technological change on the future. "Like Y2K, climate change is a technology problem that resonates with millennial anxieties." I guess that would make climate change *Y2K 2.0.*

Still, there are some important differences. Government and business spent on the order of $100 billion dollars fixing Y2K, but the problem of climate change is a lot bigger and a lot harder to solve. "And ultimately, the Y2K story ends happily with a bunch of geeks saving the world from a stupid problem the geeks themselves created," Saffo says. He thinks it will take a lot more than an army of geeks to fix the climate bug. And, of course, it's not clear that the climate story will have a happy ending.

As a technology forecaster, Saffo helped persuade the business community to get to work quickly on Y2K. "Actually, it was pretty simple. I told them, 'This is not hype. You can either fix the bug now, or you can wait until the last minute. But the longer you wait, the more expensive it will be to fix and the tougher it will be to hire people to fix it,'" Saffo explains. By then, businesses had already been running their own tests. And the outcome of the tests, which consisted of nothing more complicated than advancing their computer clocks out in time to the year 2000, suggested that Y2K was indeed a problem. When the clocks got to the year 2000, their computers stopped working. That's what you might call a straightforward modeling experiment.

Even so, some businesses underreacted to Y2K at first, and then, just as Saffo had warned, they spent more money than they should have scrambling to fix the bug in their software. "I liked to use the sailboat analogy," Saffo explains. "I'd say, 'Imagine you've got a sailboat and you need to sail around an island. You can start to circle when you're still a mile from shore and it will be easy. But if you wait until you're only 100 meters away, there will be rocks and reefs. There will be a lot more drama.'"

But ultimately, the business sector wasn't worried about drama so much as it was lured by potential opportunity.

"What really persuaded them in the end," Saffo says, "was that we presented Y2K as an opportunity. We said, 'Don't just solve the Y2K problem; use this as an opportunity to improve your business.'" I guess in the end, we all live in the hope of a better future. The problem with climate change is that it presents a set of different futures and forces us to choose one: continue "business as usual" and live on a hot planet with rising seas or change course and rebuild our energy and transportation infrastructure. Either way, we will have to pay. And either way, Saffo's sailboat analogy still applies.

———

When you're talking about the future impact of climate change on New York City, Saffo's sailboat analogy requires some modification: just pretend you're on the island instead of in the sailboat. But you're not in one of the buildings that make up the city's famous skyline or in one of the yellow cabs that snarl traffic in the streets and avenues. Instead, you're in a rather unassuming building above a place called Tom's Restaurant.

I am convinced that being a scientist in New York City is a very special experience, because New York is an *anti*–ivory tower.

Alternate side of the street parking forces the scientists who work at the Goddard Institute for Space Studies (GISS) to dash out of seminars and move their cars: parking tickets pile up if these scientists don't pay attention to the real world. And the ground floor of GISS, one of the most important climate modeling centers in the world, is home to Tom's Restaurant, a greasy spoon and Columbia University hangout made famous by the television show *Seinfeld*. I took classes at GISS as a grad student, and I found that the smell of french fries permeates the entire building. I don't know how GISS scientists can get anything done.

It's in this setting that Cynthia Rosenzweig works. Rosenzweig, a senior research scientist who heads up the Climate Impacts Group at GISS, hopes Americans can be convinced that, as with Y2K, fixing the climate bug is an opportunity to be seized sooner rather than later. And she's spent her career proving that climate change is not hype.

Rosenzweig came to GISS as a young graduate student to work on agriculture in the early 1980s. "I arrived at a time when GISS was developing some of the first global climate model projections. Jim was the director," she says. Jim is James Hansen, a well-known climate scientist and an outspoken advocate of reducing emissions. Hansen is still the director of GISS, which still has its headquarters in a nondescript building on the corner of 112th Street and Broadway.

Rosenzweig's graduate work was in agronomy, and perhaps because she entered the climate community as somewhat of an outsider, she has a talent for connecting the mathematicians, atmospheric scientists, and physicists who know the climate change projections with the economists, policy makers, and engineers who have to figure out what to do about this issue. Rosenzweig has come up with a way to study the messy business of climate impacts and offer a range of solutions.

"As scientists, we had all done so much work on the land-based resources, the ecosystems, the agriculture. We were racing to figure out what climate change would do to our ice caps, our forests, our food supply. And I began to think we were missing something really important: we were missing what climate change would do to us," Rosenzweig says. "Over 50 percent of the world's population lives in cities. I'd lived in Manhattan almost my entire life and I began to realize that we had better find out how climate change is going to affect cities, because that's where the people are." And like a sailboat navigating a turn around an island, Rosenzweig's research began to shift from studying the impact of climate change on nature to the impact of climate change on *human nature*.

In the year 2000, Rosenzweig was tapped to lead the Metropolitan East Coast Assessment,[2] one of eighteen research projects that came to be known as the National Assessment. The goal of each regional assessment was to understand the impact of climate change on infrastructure and people. Rosenzweig, also a coordinating lead author of the IPCC Fourth Assessment Report, was able to apply that experience to the study of climate impact being done for New York. And in August 2008, much to Rosenzweig's delight, the Metro East Coast Assessment led to the creation of the New York City Panel on Climate Change (NPCC), modeled on the IPCC. "New York City has its very own IPCC," she explains. "There is no other city in the world that can say that."

Cities cover less than 1 percent of the Earth's surface, but they hold half the population and produce about 70 percent of the total greenhouse gas emissions. That's why focusing on reducing emissions in cities is so important. Actually, despite their big collective carbon footprint, the cities' reliance on mass transit and smaller, stacked living spaces makes them very energy-efficient. For example, the carbon footprint of the average New Yorker is less than one-third the size of the average American's carbon footprint.[3]

Another compelling reason to tackle global warming in cities is that they are very vulnerable to a changing climate. And when you live in a city that is also an island, you've got even bigger worries. The NPCC, using data and models to project future climate change for New York City, has identified some of the most serious potential risks to New York's infrastructure. After tallying up all the risks, the panel members hand off their assessment to the mayor and the New York City Climate Change Adaptation Task Force. It's up to them to decide what the city should do about climate change. The task force consists of thirty-eight city, state, and federal agencies; regional public authorities; and private companies that operate, maintain, or regulate critical infrastructure. And as Rosenzweig sees it, New York City has decided to fix the climate bug now. The city has decided to see climate change as an opportunity, just like Y2K.

In February 2009, the NPCC released its latest report, *Climate Risk Information*. This presents a picture of what New York could look like in the future under different scenarios for greenhouse gas emissions. The conclusion is simple: the more greenhouse gases we emit globally, the more problems New York will have locally. The report shows New York as climate scientists like Rosenzweig see it. It digs, in detail, into all the gritty infrastructure issues the city is facing: everything from sewer pipes backing up to power lines sagging to airport runways flooding. As with most things in life, the details, the fine print, will get you in the end.

The report is fascinating to read if you're an infrastructure junkie like me, and Rosenzweig admits that she has become almost obsessed. "I have been all around the world working on climate change, but it became so much more real to me when I came to grips with it here in New York," she explains. There's an old saying: what you work on works on you. And as far as New York is concerned, rising temperature, increased risk of flooding, and a rise in sea level are working to make it more vulnerable.

"For New York, climate change means blackouts. That's just pure and simple," says Steve Hammer. Hammer is the director of the Urban Energy Program at Columbia University's Center for Energy, Marine Transportation, and Public Policy. He's been working on a statewide project, looking at the impact of climate change on New York state's energy supply. You can't get around it: energy is intimately connected with temperature. And the temperature in New York is going up.

The average annual temperature in New York City from 1971 to 2000 was approximately 55°F. And if you look over the long term, you'll see an upward trend. Since 1900, the annual average temperature in New York City has risen 2.5°F. There is plenty of natural variability in the record, but nonetheless, this long-term trend is interesting. And when you look at summer heat, it gets even more interesting. From 1971 to 2000, New York City averaged about fourteen days a year with temperatures over 90°F, and there were about two heat waves a summer. A bona fide heat wave is defined as three or more consecutive days with maximum temperatures above 90°F. A 100°F day in New York is actually a rarity: less than one day per year hits that level.

Of course, the number of extreme events in any given year varies a lot. For example, in 2002 New York City experienced temperatures of 90°F or higher on 33 days. But two years later, in 2004, there were only two days of 90°F or higher. What's interesting is that seven of the ten years with the most days over 90°F have occurred since 1980. Climate models suggest that the frequency and duration of heat waves will continue to increase unless greenhouse gas emissions are sharply reduced. You can also use climate models to estimate how the statistics of hot days will change in the future, on the basis of different emissions scenarios. Even with the scenario involving modest greenhouse gas emissions, known as A1B (see the accompanying table), high temperatures will steadily go up, forcing people to turn up the air-conditioning. By the end of the century, 100°F days will no longer be such a rarity.[4]

That's why Hammer, as someone who makes policy regarding energy, is worried about climate change.

"Energy systems are generally rated for a certain temperature and power load. If you keep running your power plant full blast for ten days during a heat wave, that's when things begin to break down. We know AC demand is going up and we know heat extremes are

| Temperature | 1971–2000 | 2020s | 2050s | 2080s |
|---|---|---|---|---|
| 90°F | 14 | 23–29 | 29–45 | 37–64 |
| 100°F | 0.4 | 0.6–1 | 1–4 | 2–9 |

Number of hot days in New York over four different time periods, based on observed data and an ensemble of sixteen climate models and three emissions scenarios. These values represent the central range (67 percent) of the model output. (SOURCE: NEW YORK CITY PANEL ON CLIMATE CHANGE)

going up. That's why the city is extraordinarily vulnerable," says Hammer.

And as the number of hot days begins to increase, materials begin to break down: concrete, bridges, rail lines. More and more strain is placed on the materials that make up New York's extensive infrastructure.

"For example, the capacity to transmit electricity over power lines drops with higher temperature due to increased resistance," explains Hammer.

The NPCC looked at temperature projections over the coming century. The analysts used sixteen climate models and three emissions scenarios—each scenario assuming a different human reality in the future. The first emissions scenario describes a world with rapid population growth and limited sharing of technology. The second emissions scenario describes a world where the effects of economic growth are partially offset by the introduction of new technologies and decreases in global population after 2050. The third emissions scenario describes a world where global population grows to about 9 billion by 2050 but then declines to about 7 billion by the end of the century. This is also a scenario in which society places emphasis on clean, efficient technologies that reduce the growth of greenhouse gas emissions. As a result, it has the lowest greenhouse gas emissions of the three, with emissions beginning to decrease by 2040.

Keep in mind that the range of temperatures predicted for New

York is mostly a reflection of these different emissions scenarios. If emissions are kept down, New York will be likely to stay along the lower end of the temperature range. But if nothing is done to reduce emissions, New York will be likely to see temperatures that are even higher. According to the NPCC, New York's average temperature is expected to increase by about 1.5°F to 3°F by the 2020s, 3°F to 5°F by the 2050s, and 4°F to 7.5°F by the 2080s. In other words, by 2080, the overall climate of New York City will be more like that of Raleigh, North Carolina, or Norfolk, Virginia, if greenhouse gas emissions aren't sharply reduced in the coming years.

Although this may sound good to people who love the heat, the problem is that New York's energy infrastructure wasn't built with Raleigh's climate in mind.

"The energy risk becomes very apparent when you look at what percentage of the overall power load in the state is currently going toward air-conditioning," says Hammer. "Not too long ago, that number represented approximately 2 percent of total electricity load. But the thing is, turning on the AC creates peak power demand problems. As that demand increases, you'll need to dramatically increase the amount of total power generation capacity available around the state. You can't expect to satisfy peak demand increases by drawing in power from other places, as demand increases will be a regional phenomenon," says Hammer.

If they're not made more energy efficient, cities have the potential to become trapped in a vicious circle with regard to climate. More heat extremes lead to increased energy demand, which leads to more heat extremes. That's why fixing the climate bug is so crucial.

There are two ways to do it. One way is called *mitigation*: you can fix the climate bug by reducing greenhouse gas emissions. Another way is known as *adaptation*: you can cope with the climate problem by fixing the infrastructure. Rosenzweig and the rest of the panel on climate change are trying to show New York how to do both. "We're trying to create this road map whereby climate information can bear on other areas of society," Rosenzweig says. The thinking is that if people can see what New York might look like in the future, they will opt to avoid the unnecessary drama, of which rising temperature is just one example.

"I know it's old-fashioned. But I am very much into win-win solutions. And anything we do today will help us today," Rosenzweig adds.

That ended up being true of Y2K as well. Saffo says you can credit the millennium bug for the swift rebound of New York City's computing systems after the attacks of 9/11. "Y2K forced Wall Street to make upgrades. Wall Street had a Y2K drill. They practiced that drill and it paid off," Saffo says. The system redundancies developed in anticipation of Y2K allowed the city's transportation and telecommunications sectors to provide service despite the enormous damage on 9/11. Those redundant networks and contingency plans put in place by an army of geeks led to an opportunity that may well have saved lives.

"And guess what? We are *not* perfectly adapted to the climate extremes of today!" says Rosenzweig. "That's why everything we do to adapt now is going to help us right now."

The power grid isn't the only area that remains unsuited to the climate extremes of today. Flood protection is another glaring example that Rosenzweig cites to show how ill-suited we are for the present, let alone the future.

"The Saw Mill River Parkway"—a major traffic artery that connects New York City with the suburbs to its north—"floods now every time there's a heavy rain," says Rosenzweig. "So why don't we, as a region get organized and realize, Hey, these types of events are going to happen more frequently. There is no need for everyone to pile onto the parkway and sit in these big puddles. We're smarter than that. It could be as simple as having everyone telecommute that day." It turns out that not all adaptations are expensive, and not all involve infrastructure. A lot of the issue is just how we choose to manage and operate the infrastructure.

Rosenzweig is really worried about flooding. "Without a doubt,

New York's biggest vulnerability is enhanced coastal flooding due to sea level rise. The sea level rise is guaranteed. It's unidirectional. Just like the warming," says Rosenzweig. When you live on an island, a little sea level rise goes a long way. But this isn't a problem just for the people who live along Manhattan's coastline. A lot of New York City's critical infrastructure sits less than 10 feet above mean sea level; and experts like Rae Zimmerman, a member of the NPCC and a professor of planning and public administration at New York University, say that anything below 10 feet is vulnerable to flooding during major storm events.[5]

If you look at Zimmerman's list of vulnerable transportation infrastructure, you'll see dozens of well-known places, including the Canal Street Subway Station in Chinatown, which is 8.7 feet above sea level; the Christopher Street Subway Station in Greenwich Village, which is almost 15 feet below sea level; the New York entrance to the Holland Tunnel, at 9.5 feet above sea level; and LaGuardia Airport, at only 6.8 feet above sea level. (See Appendix 2 for more details.)

Until about 150 years ago, sea level had been rising along the east coast of the United States at a rate of about 0.34 to 0.43 inch per decade. It had been rising at this rate since the end of the last ice age, mostly because of regional subsidence or sinking, as the Earth's crust still slowly readjusted to the melting of the ice sheets. But within the past 150 years, as global temperatures have increased, regional sea level has been rising more rapidly. At present, rates of rise in the sea level in New York City range between 0.86 inch and 1.5 inches per decade. The long-term average rate since 1900 is 1.2 inches per decade. In Lower Manhattan, the water at the Battery has risen more than 1 foot during the last century.

As a result, the 1-in-100-years flood, or the flood that has a 1 percent chance of occurring in any given year, will happen about every eighty years instead. The current 1-in-100-years flood can produce a sea surge of approximately 8.6 feet for much of New York City, and that surge height is shifting upward just as the chance of such a flood is shifting upward. Flood statistics in general are changing. By the end of the twenty-first century, the kind of coastal floods that at present occur about once per decade may occur every other year. According to the World Bank Climate Resilient Cities report, a 100-year flood may increase from once in eighty years, where it is today, to once in forty-three years by the 2020s and to once in nineteen years by the 2050s.

"This goes beyond the climate forecasts," says Rosenzweig. "This is fundamentally about our values. This is about who we are as a city." Rosenzweig has already been a host for visiting Dutch flood control experts. "We have a lot to learn from the Dutch about adaptation planning," she says. The Netherlands has waged a long battle against the sea, and this struggle has clearly shaped Dutch values. The Dutch have lost enough battles to be very concerned about climate change.

One of the Netherlands' biggest battles came on the evening of January 31, 1953, when a high-tide storm breached its famous dikes in more than 450 places. You could argue that this flood reshaped the political, environmental, and psychological landscape of the nation as much as it reshaped the land. More than 1,800 people died, many as they slept. More than 47,000 homes and buildings were swept away.

Twenty days after that devastating flood, the Delta Plan was launched. Dutch politicians set in motion a $3 billion, thirty-year program to end the threat from the sea once and for all. The Dutch decided that their strongest sea defenses would be designed to stand up against a storm so strong it would occur only once in 10,000 years. The river levee and dike systems were built to withstand a 1,250-year storm. The country built an elaborate network of dikes, man-made islands, and a 1.5-mile stretch of sixty-two gates to control the entry and exit of North Sea waters into and out of the low-lying southwestern provinces. The Delta Plan is one of the largest construction efforts in human history and is considered by the

American Society of Civil Engineers (ASCE) as one of the seven wonders of the modern world.

New York—like the rest of the United States—doesn't get nearly that kind of praise from the ASCE. In fact, in its 2009 Infrastructure Report Card, the ASCE gives America's total infrastructure a D. In New York State, ASCE's most serious concern is bridges, roads, and mass transit. The engineers found that 46 percent of New York's major roads are in poor or mediocre condition, 42 percent of New York's bridges are structurally deficient or functionally obsolete, and 45 percent of New York's major urban highways are congested. In addition, there are 391 high-hazard dams in New York. A *high-hazard dam* is defined as one whose failure would cause loss of life and significant property damage. Forty-eight of New York's 5,089 dams are also in need of rehabilitation to meet the state's applicable safety standards. One explanation for all this is that by 2030, just about all of New York's major infrastructure networks will be more than a century old.

As problematic as the dams and bridges are, they're only part of New York City's infrastructure problem. The city's first subway line opened in 1904, the same year as the first New Year's Eve bash in Times Square, and the subway signaling technology that's still in use today was built before World War II. The energy grid was started in the 1880s, and two of the city's water tunnels were completed before 1936. (A third is being built now and is expected to be finished in 2020.)

In short, New York is an old city facing new problems. In April 2007, the city launched a comprehensive sustainability plan: PlaNYC 2030.[6] The 2007 plan now covers both mitigation and adaptation. New York is fighting its own battle with the sea—on two fronts.

In New York, the big storms come in two types: hurricanes and northeasters. Hurricanes are more likely to cause the 1-in-100-years and the 1-in-500-years floods. Northeasters, the ferocious winter storms named for the continuous, strong northeasterly winds blowing cold air down from the Arctic, are the main source of the 1-in-10-year coastal floods.

"Tell me when the next hurricane's going to hit and I'll tell you how soon we have a big problem in New York," says Hammer. As with the Dutch in 1953, predicting when the next big hurricane will hit Manhattan is impossible. All you can do is build, knowing it's going to happen again at some point.

Historically, hurricanes have hit New York (see Appendix 2) between July and October. On the basis of this short record, the National Hurricane Center estimates the *return period* for a category 1 hurricane in New York at about once every twenty years. A category 1 hurricane has sustained winds of between 74 and 95 miles per hour on the Saffir-Simpson hurricane wind scale. A return period of twenty years for a category 1 hurricane at a given location means that on average during the previous 100 years, a category 1 or greater hurricane has passed within 86 miles of this location about five times. The return period for a category 3 hurricane, which has sustained winds of 111 to 130 miles per hour, is roughly once in seventy years for New York. To put this period in perspective, it's about one in ten for Miami.

According to a 1995 study by the U.S. Army Corps of Engineers, a category 3 hurricane in New York could create a surge of up to 16 feet at LaGuardia Airport, 21 feet at the Lincoln Tunnel entrance, 24 feet at the Battery Tunnel, and 25 feet at John F. Kennedy International Airport. And that's with sea level measurements as of 1995. The impact could be even greater if the storm hit at high tide, as was the case in the Netherlands in 1953. The Army Corps of Engineers estimates that as many as 3 million people would need to be evacuated from New York City.

"The Dutch have decided a 1-in-10,000 year standard is right for them," says Rosenzweig. "And we have to focus on doing what's right for New York. But we need to decide now." A category 3 hurricane is capable of producing sustained winds of more than 111 miles per hour, but the current building code requires windows to withstand only gusts of 110 miles an hour. Add to that the fact that wetlands in and around New York City, the natural sponges that help absorb some of the damage done by hurricanes, have shrunk by almost 90 percent over the past century, owing to a combination of development and storm damage. Numerous decisions need to be made. For example, some scientists have suggested that New Yorkers follow the lead of the Dutch and construct storm surge barriers. These experts have simulated storm surge to evaluate the effectiveness of barriers at several points in New York Harbor. Two historical storms were evaluated—Hurricane Floyd and the northeaster of December 1992—and in both simulations the barriers were shown to be operationally effective. But such barriers are very expensive and won't provide protection everywhere. The decision is up to New York.

———

Hurricanes illustrate some of the most dramatic risks to New York's infrastructure, but there are plenty of everyday weather events that will become increasingly problematic for the city's support systems—most notably the guts of New York, its sewer system. New York City's drainage and wastewater system is extensive. It consists of about 6,600 miles of sewers, 130,000 catch basins, almost 100 pumping stations, and fourteen water pollution control plants.[7] The sewer system was designed to minimize standing water on roadways and streets, is mostly gravity based, and has been built over hundreds of years. The sunk-cost investment in the city's sewer systems is enormous, and the problem is that there's almost no flexibility to modify existing piping to install bigger pipes.

Rae Zimmerman is not only a member of NPCC but also the director of the Institute for Civil Infrastructure Systems at New York University's Robert F. Wagner Graduate School of Public Service. "Sewage water and storm water use the same pipes. They do it for economics, but it makes them more vulnerable in the event that something happens," she explains. In addition to the cost and disruption, the time to effect changes would be extremely long. More space would need to be found within the maze of subsurface utilities below the streets, and pumping might be needed in some instances to convey storm and wastewater flows. Despite these obstacles, some changes must be made to prevent flooding of streets and basements.

Like a lot of older cities in the country, New York City has a single, combined sewer system that handles sanitary waste as well as storm water. When it's not raining, sewage treatment plants can handle all the sewage and clean it up. But when it rains, the vast amount of rainwater that goes into the sewers exceeds their capacity, so some of it has to be released into the rivers untreated. If rainfall becomes more intense—as observed data and climate models suggest will happen—the sewer system could be overwhelmed. That would result in more flooding of streets and basements, and more untreated waste would enter rivers.

Rising sea level will also be a factor. "New York City has regulators. So when the tides go out, the pipes—the storm sewers—are exposed and the water flows out of them and into the river. But when the tide comes in and there's flooding, the regulators shut, or else you would have all the river water flooding New York City. Now, if there is sea level rise to the point where those regulators are always shut, there is no place for any storm water to go, and it will all spill onto the street," Zimmerman explains.

There are some simple and cheap solutions that can help alleviate the strain. Staten Island is already preserving its natural drainage corridors, called *bluebelts*. These bluebelts are nothing more than streams, ponds, and other wetland areas that act to convey, store, and filter storm water, with the added benefit of providing open spaces and diverse wildlife habitats. Bluebelts have been shown to save tens of millions of dollars in infrastructure costs, compared with conventional sewers. And in Hendrix Creek, a tributary to Jamaica Bay, ribbed mussel beds have been reintroduced to test this mollusk's ability to improve the water quality of tributaries around combined sewer overflow outfalls.

As always with climate change, there is the problem of too much water and too little water. Approximately 90 percent of the city's water supply is from the Catskill and Delaware systems. New York City's drinking water originates from a watershed about 2 square miles in area, situated 125 miles north of the city. It provides 1.1 billion gallons per day to 8.2 million city residents plus an additional 1 million people upstate. The system has a network of nineteen reservoirs and three controlled lakes throughout the Croton watershed east of the Hudson River and the Catskill and Delaware watersheds west of the Hudson. Some of the new problems associated with climate change could compromise the existing water supply and treatment systems.

As the temperature increases, more precipitation will fall as rain than as snow; consequently, there will be less storage and therefore reduced inflows to reservoirs during the spring season. Peak snowmelt in the Catskills is already shifting to earlier in the year. A recent study shows that during the period from 1952 to 2005 it shifted from early April to late March. In this scenario, the juggling act between floods and water supply gets harder. Lower reservoir levels can protect against a sudden flood, but low levels may reduce the statistical probability that water resources will be available to the city. A drought watch is declared when there is no better than a 50 percent probability that either the Catskill or the Delaware reservoir system will be filled by the following June 1. This definition is based on records going back to 1927—but such records are not an accurate predictor of the future.

New York uses about 1,060 million gallons of water per day (mgd). (In more familiar terms, this is 1.06 billion.) But demand can easily rise to more than 2,000 mgd (2 billion) during heat waves. For example, August 2, 2006, was the third successive day with temperatures in the 90s and humidity over 70 percent. The daily flow was 1,560 mgd (1.56 billion), but the peak flow reached

2,020 mgd (2.02 billion). During a heat wave in the city, illegal hydrant usage jumps up, just at a time when water demand is at its highest. And the ability of the existing aqueducts to refill the city's main distribution reservoir is pushed to the limit. You can usually expect water levels to go down and water pressures throughout the entire system to become quite weak. That spells trouble for firefighters, increases the number of complaints about low water pressure on the upper floors of buildings, and increases sediment resuspension within water mains. You can almost hear the city groaning under the strain.

––––––

For all the drama it is capable of creating, climate change is ultimately about a million boring little fixes. But as Rosenzweig notes, these boring little fixes can have a profound impact. One actual example of an adaptation strategy is installing more fire hydrant locks. And in May 2008, the Climate Action and Assessment Plan published by the Department of Environmental Protection called for the installation of better hydrant-locking mechanisms.

More frequent droughts will begin to work in combination with rising sea level in ways that will affect the water supply. "All these systems are intertwined," Rosenzweig explains. "Climate change impacts in urban areas like New York are completely integrated. That's why our science teams have to be integrated. If they're not, we get the wrong answer." In New York, it is essential to take a multidisciplinary approach that involves scientists with many different backgrounds, because they can all come at the problem from a different point of view.

As Rosenzweig explains, "We meet every month with our full team of scientists. The hydrologists tend to be very confident. They often say, 'Don't worry, we can handle it, climate variability is our middle name.' And so we were discussing the fact that the Palmer Drought Index, a commonly used index used to measure drought severity, shows more frequent droughts in the New York region.

The hydrologists said, 'Don't worry, we have a pipe that goes into the Hudson River at Chelsea, which is a little town up the river, so we're just going to supplement the supply by taking in water from the Hudson River at Chelsea.' And a colleague from the NPCC waves her hand in the back of the room and says, 'But we just calculated that when you include the impacts of sea level rise, the salt front in the tidal estuary would swing all the way up to Chelsea.'

"This is a classic example of why we need all these experts in the same room. Sure, we can take in additional water from the Hudson River at Chelsea. But if it's salt water, that's not going to help us very much."

PlaNYC has set a goal to reduce New York City's greenhouse gas emissions 30 percent by 2030. There are four principal strategies in the mitigation plan: avoid sprawl, generate clean power, make buildings more energy efficient, and create sustainable transportation. New York City's population is expected to grow from 8.36 million today to about 9.1 million by 2030. Scientists agree that far deeper emissions reductions, on the order of 80 percent, will be necessary by 2050 if we are to stabilize global temperatures. In New York, the reduction plan will focus mainly on buildings, which contribute 80 percent of greenhouse gas emissions, compared with about 35 percent nationally. And 85 percent of the 950,000 buildings in New York City in 2030 already exist today. The city has promised to do a carbon inventory every year to track its progress.

Zimmerman likes to point out that New York shouldn't focus solely on climate change. "I am a firm believer in the fact that we can be addressing national security, climate change, and sustainability simultaneously. In fact, I don't think we should be doing any of this stuff separately," she says. Zimmerman would like to see the city diversify and decentralize a lot of its infrastructure. "If we generated more electricity from solar and wind, we'd be reducing greenhouse gas emissions and we'd also be reducing the consequences of terrorist attacks because the production of energy would be more decentralized. The goal is to have a building-by-building supply," she says. "I believe in decentralized infrastructure, but that doesn't mean decentralized cities. I believe in cities!"

In keeping with this building-by-building approach, the rooftops of New York City hold a lot of hidden potential. One study cited by PlaNYC calculates that if all the rooftops in the city were covered with solar panels, they could produce nearly 18 percent of the city's energy needs during daytime hours. In New York, roughly 40 percent of the carbon footprint comes from electricity consumed, and another 40 percent comes from the heating fuels burned directly in buildings. So making buildings more efficient is a major part of the strategy to reduce carbon emissions. It can also reduce air pollution. Nearly one-third of locally produced particulate matter in our air comes from heating fuel. Public health can improve quickly as a result of efforts that improve air quality and building efficiency.

New York may be looking to learn from others, but it's also hoping to serve as an example of best practices. Steve Hammer also serves as an adviser to the Energy Smart Cities mayoral training program being developed by the Joint U.S.–China Cooperation on Clean Energy, a nonprofit with offices in Shanghai and Beijing. The training program is a three-year initiative that will introduce Chinese mayors to the best practices of the West, with the goal of helping cities reduce their energy intensity by 20 percent by 2010. Energy intensity, the ratio of energy use to output, is a way to measure the overall energy efficiency of an economy. Hammer brought Dr. Rit Aggarwala, the lead architect of the PlaNYC report, to China to speak at the first training session, along with a dozen other international experts.

In general, Hammer says, climate change is getting much more attention in China. "About four years ago, there was one wind turbine blade manufacturer that had set up shop in China. Today there are over seventy making blades and other turbine parts. The market is shifting very rapidly here," says Hammer. But for all this market growth, Hammer acknowledges there is much to be done at the local level. "The mayors I've met generally agree they can do a better job at local sustainability matters. Many would like to take this on, but they need help in identifying strategies that are appropriate for their city. China's central government could help prime this pump by investing in local energy planning, much like the Obama administration has started to do in the United States."

Swift action is essential, as the rate of urban growth in China is dramatic. "The country will build 50,000 new skyscrapers in the next twenty years, and what could be more important than making them sustainable? In 1980, Shanghai had 112 buildings more than eight stories high. Now it has 13,000 of them," says Hammer. "It's like an almost infinite Manhattan."

"The question I get asked the most," Rosenzweig says, "is: am I optimistic or pessimistic about the future? I really feel climate change is the issue that is challenging human beings to become sustainable. It's an issue that is finally big enough and destructive enough that it forces us to pay attention. And if sustainability becomes a full-fledged universal value, we'll be OK." Rosenzweig, even knowing everything she knows, is optimistic.

Paul Saffo is not. He says we're still stuck in a debate between two camps over what exactly to do about climate change and sustainability. "On the one side you have the engineers and on the other side you have the Druids. I love them both," he says. "The problem is that the engineers say let's fix our way out of this; let's use technology and go faster in the future. The Druids say let's go slower. Let's go back to a time when things were smaller and simpler. The climatologists can't seem to decide which they are."

But maybe that's the point. Maybe the climatologists know we have to do both. We have to look forward and we have to look back.

In 1609, 400 years ago, the English explorer Henry Hudson sailed into New York Bay. Hudson had been hired by the Dutch East India Company to find a northeast all-water trade route to Asia. On his ship, the *Half Moon*, Hudson and his crew left Amsterdam and sailed along the coast of Norway until they hit ferocious weather and sea ice. Rather than return empty-handed, and disobeying orders to search only for a northeast route, Hudson made a 3,000-mile detour in search of warmer weather and the dream of finding a southwest route to Asia through North America. He never found the southwest route, but he did make another discovery, the island of Manhattan. The Dutch colonial settlement and fur trading outpost of New Amsterdam served as the capital of New Netherland until it later fell under British rule and was renamed New York. New York wasn't the dream Hudson had been chasing, but it has come to represent the hopes and dreams of millions.

"I guess," says Rosenzweig, "everyone has the New York of their dreams. For me, it's a climate-resilient city." The year 2009 marked the second anniversary of PlaNYC.[8] On Earth Day, April 22, Mayor Bloomberg announced that eighty-five of the plan's 127 initiatives were on time or ahead of schedule. As part of PlaNYC, the city had converted 15 percent of the yellow taxi fleet to hybrid vehicles; planted 174,189 trees across the five boroughs; acquired 13,500 acres of land to protect the upstate water supply; saved 327 tons of nitrogen oxide ($NO_x$) per year by means of retrofits to the Staten Island ferry fleet; and started twenty storm water retention pilot projects. Maybe Rosenzweig is right to be optimistic.

## New York, New York: The Forty-Year Forecast— Hurricanes, Infrastructure, and Sea-Level Rise

| NEW YORK, NEW YORK | TODAY | | 2050 | | 2090 | |
|---|---|---|---|---|---|---|
| Emissions Scenario | JAN. | JULY | JAN. | JULY | JAN. | JULY |
| Higher | 31.6 | 74.7 | 34.9 | 78.1 | 38.4 | 81.9 |
| Lower | | | 34 | 77.4 | 35.4 | 78.6 |

## Forecast
## September 2013

You could say New York deserved a lucky break. It had already been slapped around enough by tropical storms. Back in September 2004, the remnants of Hurricane Frances had flooded its subways and stranded passengers. And in September 1999, Hurricane Floyd, by then weakened to a tropical storm, had dumped more than 10 inches of rain on the city, causing mudslides on the bluffs overlooking the Hudson River near the Tappan Zee Bridge. There were still plenty of people who remembered Hurricane Donna, the category 3 storm that pounded New York City on September 12, 1960, with sustained winds of more than 90 miles per hour. Donna had flooded lower Manhattan almost to waist level on West and Cortlandt Streets—the southwest corner of what later became the site of the World Trade Center. The last thing New York needed was another hurricane.

An average hurricane season has eleven named storms and six hurricanes, including two major hurricanes. The United States Landfalling Hurricane Probability Project put the risk that New York would be hit by a major hurricane (category 3 or more) by 2050 at 90 percent.[9] We all knew that eventually, with or without global warming, a major hurricane was going to hit New York. The question was simply when the Atlantic Ocean would start to play hardball again.

The period from 2009 to 2012 was a stretch with the fewest named storms and hurricanes since 1997—thanks, in part, to an El Niño in 2009–2010. El Niño produced strong wind shear across the tropical Atlantic, which meant fewer and shorter-lived storms. It almost seemed as though, after producing Katrina in 2005, the Atlantic Ocean had gone into semiretirement. If hurricane seasons were anything like baseball, then the Atlantic seemed to be in a very welcome slump. For four years in a row, no major hurricanes had hit the United States. And as for baseball: the Yankees, in their new stadium, went on a winning spree of four World Series in a row.

But all good things must come to an end. And in September 2013, with the Yankees not even looking to play in the postseason, the Atlantic Ocean reawakened and one specific hurricane seemed to be in a New York state of mind. After beginning as a garden-variety low-pressure system moving off the coast of west Africa, the storm that eventually came to be known as Hurricane Homer gathered strength as it crossed over unusually warm tropical Atlantic waters. The warm ocean water acted like a heat pump, fueling the hurricane and causing it to increase in intensity. Many models still struggled to predict exactly how climate change would affect hurricanes, but there was general agreement that warmer water meant more intense storms.

The NOAA GOES-12 satellite recorded the storm's every move, and initially the National Hurricane Center issued a watch for Miami, expecting the storm to hit there. Miami's residents stockpiled supplies, boarded windows, and secured boats—as usual. But then Homer took a turn north and surprised everyone when it started speeding up. The Bermuda High, a large area of high pressure in the Atlantic, pushed the storm up the coast as warm water provided fuel to the system. In time, the hurricane would achieve a record-breaking forward speed of 75 miles per hour. As the system raced up the coastline of the Carolinas, the revised track forecast issued by the National Hurricane Center warned of a category 3 storm making its way directly toward New York City. Here was a system that bore a striking resemblance to an event that had almost happened back in 1938, when the Long Island Express, which ultimately missed New York City by just 75 miles, did tremendous damage up and down the northeast coast. The only difference was that this one didn't look as though it was going to miss.

The models suggested that New York was about to get swallowed by storm surge. Surge levels for Hurricane Homer had been calculated by the U.S. Army Corps of Engineers using NOAA's

SLOSH model, and in a worst-case scenario Homer was likely to create a surge of up to 25 feet at John F. Kennedy Airport, 21 feet at the Lincoln Tunnel entrance, 24 feet at the Battery, and 16 feet at LaGuardia Airport. The U.S. Army Corps of Engineers estimated that nearly 30 percent of the south side of Manhattan would be flooded. The storm surge flooding would threaten billions of dollars of property. Rising sea level was already a factor, as each seemingly small increase in sea level gave the hurricane a longer, more destructive reach into the city. Since 1900, sea level in New York City had been rising at rate of about 1.2 inches per decade. Hurricane Homer plus this rise meant more storm-related coastal flooding, more inundation of wetlands, more structural damage, and more money lost.

Because of the projected track and the fact that the highest, most destructive winds lay to the right of the storm's eye, it was anticipated that Homer could pass directly over New York with gusts over 150 miles per hour—shattering the glass in skyscrapers and sending razor-sharp shards raining to the ground. In addition, the counterclockwise, westerly flowing wind would funnel the surge waters into New York City harbor.

As the storm swept up the coast, we all became experts on the history of hurricanes in New York. We knew that there had been only two honest-to-goodness direct hits in New York City in recent history—the Great September Gale of 1815, and a storm that came on September 3, 1821, and made landfall at Jamaica Bay. Both were category 3 storms and both did extensive damage. With widespread flooding in lower Manhattan as far north as Canal Street, the 1821 hurricane set the record for the highest storm surge in Manhattan—nearly 13 feet. One question was whether Hurricane Homer would go down in the history books as the third direct hit. Another question was what, if anything, could New Yorkers do to prepare themselves.

For the most part, people simply jumped ship and left the city. It ended up as the largest peacetime evacuation since Hurricane Floyd. As New Yorkers headed through bridges and tunnels, transit workers stayed behind, trying to ready pumps and prep storms drains. It felt rather like rearranging deck chairs on the *Titanic*, but there was little else to do. There was simply no infrastructure to deal with a storm of this magnitude. People did what they could and then simply sat back, prayed, and vowed that they wouldn't allow the city to be this vulnerable again.

Fortunately, the Bermuda High shifted, and the hurricane began to shift course: the center of the storm headed out to the open sea. And New York, which could have been swallowed up whole and spat out by this storm, got off relatively easy.

That said, there was still significant damage to sift through. There was severe coastal erosion and heavy street flooding. The Rockaway beaches nearly vanished, owing to the high winds and storm surge. The sewer system was completely unprepared for the volume of rain that fell in such a short time. The water began pooling at the street corners, and then gradually rose to inflict damage on parked cars and storefronts. Many owners of the city's famous facades had boarded up their windows to protect the glass from the wind, but the water was one thing that no one could have prepared for.

Though no buildings were submerged, and even though the storm was a near miss, the water inflicted excessive damage across the southern third of the island. Many people who lived in basement apartments returned to the city to find their possessions soaking wet and now had to contend with an unfortunate and unhealthy problem: mold.

On top of all this, September set a new record in New York for warmth. In general, temperatures usually associated with August were now stretching deeper and deeper into September; and now New York was in the grip of a late summer heat wave. With electricity out in many parts of the city and nighttime lows still in the 80s, people felt overwhelmed by weather.

Trees were yet another casualty. Tree-lined blocks from the Upper West Side to Park Slope, Brooklyn, were stripped of their greenery. Picturesque streets suddenly looked as if they had encountered a wood chipper, with the pavement and cars covered in brown and green shrapnel. Trees were down all over Central Park; the cleanup and replanting would eventually cost several million dollars. The parts of the park that were hit hardest were closed to the public for several months, making for an unusually quiet Central Park in the fall.

And then there was the subway system. Storm tides overtopped some of the region's seawalls for only a few hours, but they still managed to flood the subway as well as the PATH train systems at the station in Hoboken, New Jersey, shutting down these transportation systems for almost a week. This shutdown made it very difficult for those who had left the city to get back home and start the cleanup. The stations that everyone knew were vulnerable to flooding in an extreme weather event proved to be just that. Because of its proximity to the Hudson River and its depth below sea level, the Christopher Street subway station was shut down longest, with water covering the tracks for several weeks after the hurricane.

Despite the tens of millions of dollars in damage, New Yorkers realized how lucky they were. But when they slowly began to file back into the city, their questions followed them home. How could a major metropolitan area like New York be so vulnerable? Why wasn't more being done to replace century-old infrastructure? If the Yankees could build a new stadium for $1.3 billion, why couldn't the mayor and the Metropolitan Transit Authority make basic improvements to the subway? After all, the Dutch and the British had already spent billions fortifying their cities against the growing threat of storms and rising sea level.

## January 2014

The high-resolution model projections of what *could* have happened to New York got under everyone's skin. Deep down, we all knew it was only a matter of time before something really happened. The city, with its elderly infrastructure and vulnerable coast, needed help. Other cities were already adapting to climate change. Boston had elevated a sewage treatment plant to keep it from being flooded; this project was based on projections from scientists at Harvard regarding the rise in sea level. And Chicago, through its Green Roof and Cool Grants Program, encouraged rooftop gardens and reflective roofs to help keep the temperature down and ease heat waves. New York, of course, had its own plans in place, and now the hurricane that almost was had given everyone a new sense of what would eventually come to be.

Organizations and volunteers across the city began to implement strategies that had been laid out by the New York City Climate Change Adaptation Task Force. Given the climate change projections performed by Columbia University's Center for Climate Systems Research and NASA's Goddard Institute for Space Studies, we understood that the city was probably headed for a 3°F to 5°F increase in temperature, a 2.5 to 7.5 percent increase in precipitation, and a 6- to 12-inch rise in sea level by 2050. We worked to rebuild the city with that climate in mind.

And so began a coordinated plan to adapt the city's roads, bridges, and tunnels; its mass-transit network; its water and sewer systems; its gas and steam production and distribution systems; its telecommunication networks; and other critical infrastructure to be able to deal with the likelihood of more extreme weather. There were plans to modify dam infrastructure and allow for water releases in anticipation of a storm, and to inventory existing tide gates and identify priority locations most vulnerable to a rise in sea level and to storm surges. There were programs that focused on the long-term viability of New York's water and sewer systems. There were land acquisition programs aimed at protecting the watershed, as well as a plan for new water quality infrastructure. A complete face-lift for the city would cost an unthinkable sum. So the work was done piecemeal and at the local level by concerned groups of public and private organizations. I'm not really sure how long a New York minute is, but we were now thinking about a New York century. It was as if the entire city had become obsessed with adapting to climate change. Go figure.

## April 2017

It was a good thing that we did begin to chip away at the city's infrastructure shortcomings. In April 2017, four freak storms each dumped more than 7.5 inches of rain in upper Manhattan in one day, setting new records for daily accumulation. Another three days saw between 4 inches and 6 inches of rain at locations across the city within a four-hour period. In some areas, 4 inches of rain fell in one hour alone. The city's existing storm water conveyance system had been designed for only 1.75 inches of rainfall per hour. We added "new storm drains" to the city's wish list. Needless to say, the April storms flooded the subway, too. It was obvious that we also needed to focus our attention underground.

There were at least three situations that could take out the subways: (1) flooding of the tracks over the third rail; (2) water pouring through street-level vents, leading to a smoke condition; and (3) flood impacts on the signaling system. As rainwater seeped through tunnel walls and headed down subway grates and stairwells, sump pumps in 280 pump rooms next to the subway tracks would pull the water back up to street level.[10] That water then naturally flowed toward the storm drains. But more often than not, the storm drains were unable to handle the flow of water. They were designed to take away only so much water. Eventually, the water would make its way onto the subway tracks, hitting the all-important third rail. The 600 volts of electricity running through the third rail would cause the water to boil and set all the floating debris on fire. In addition, the water would short-circuit the electrical signals and switches, making it impossible for train operators to know when it was safe to stop or go. The subway system had two critical sources of power: the direct current propelling the trains, and the alternating current powering the signals. Even if the direct current was spared, the trains couldn't run without signals.

It took time, but the New York City Transit Authority installed additional pumps and built new storage tanks for processing rainwater runoff. It also developed a computer system to better monitor storage during times of overflow. Also, the Department of Environmental Protection moved electrical equipment, such as pump motors and circuit breakers at the Rockaway Wastewater Treatment Plant in Queens, from 25 feet below sea level to 14 feet above sea level. The beach nourishment projects that had been going on since 1924 to prevent coastal erosion were also stepped up in the Rockaways as a way of dealing with the effects of a rising sea level. Depositing more sand on the beach strengthened the defense that the beach provided during coastal flooding.

But in the long term, it became increasingly clear that a new network of storm surge barriers would need to be constructed in order to protect the city. As the sea level rose, New York became even more vulnerable to storm surge flooding. It would take high-water levels of only 4.9 to 5.7 feet above mean sea level to cause flooding over some of the southern Manhattan seawalls. Global warming was expected to increase the rate at which sea level rose: from about 1 foot per century to between 1.6 and 2.5 feet per century.

It was suggested that four barriers 30 feet high and able to withstand a 1,000-year flood event should be constructed: two off the coast of Staten Island, one to the northeast and one to the southwest; one off the coast of the Bronx, protecting LaGuardia Airport; and one off Breezy Point in the Far Rockaways to help protect John F. Kennedy Airport. The barriers would need to operate with as little as one hour's warning, closing each gate in fifteen minutes and the complete barrier in thirty minutes. The construction time was estimated to be eight years.

## July 2027

As the models projected, temperatures across the city had been steadily creeping up. The daily highs were getting higher, and so were the nightly lows. By the 2020s everyone realized that summers were longer and hotter than ever before. That's why it came as no big surprise when New York broke the record for the longest stretch of 90-degree days ever recorded. For twenty-one days straight, New York's aging power grid was brought to its knees. The increase in peak electricity load when people cranked up their air conditioners resulted in routine brownouts as well as an increase in costs associated with cooling water for power plant operations. Even with the broad mandate for infrastructural repair that had been handed down after Hurricane Homer, the grid proved too complex to overhaul. Unfortunately, this delay had deadly consequences, which began with a lightning strike.

The lightning strike that set off the New York blackout of 2027 caused the loss of two transmission lines and a subsequent loss of power from the nuclear plant at Indian Point. New York Power Pool Operators called for Con Edison operators to "shed load."

Because of the power failure, LaGuardia and Kennedy airports were closed for about eight hours, automobile tunnels were closed owing to lack of ventilation, and 10,000 people had to be evacuated from the subway. Con Ed called the shutdown an "act of God," enraging the politicians in City Hall, most of whom said that the utility was guilty of gross negligence for not working to better prepare the power grid for a heat wave that had been years in the making. In many neighborhoods, veterans of the Northeast blackout of 2003 headed to the streets at the first sign of darkness. But many of them did not find the same spirit. In poor neighborhoods across the city, looting and arson erupted.

## January 2039

The storm surge barriers were over budget and not even close to being on schedule, but they were finally done. This was the eight-year project that took twenty-two years to complete. Regardless of the complaints and infighting, we were all just glad to see the job finally done. And there was more good news. The runways at Kennedy, LaGuardia, and Newark airports were raised in anticipation of higher flood levels. There was also a plan in place to move the West Side Highway inland. Coastal areas were rezoned for parks and recreational uses, not high-density residential development.

None of it was cheap, and plenty of people began to worry that it was overkill. Elevating single-family homes in Long Island by 2 feet could cost anywhere from $22 to $62 per square foot. Additional seawalls cost about $5 million per mile.[11] For the country as a whole, it was estimated that building seawalls to protect the United States from coastal flooding would cost from $46 billion to $146 billion.[12] Scientists had also come up with ways to incorporate mangroves and sea grasses into the design of seawalls to improve their environmental impact and make them look better, too.

It was strange to see the climate projections playing out before our eyes. What used to be the 1-in-10-years coastal flood, fifty years ago, now came every other year. And the 1-in-100-years coastal flood happened four times more often. It was a good thing we had raised those wastewater treatment plants.

Over the years, new research had begun to suggest that the rise in sea level along New York's coast would be much higher than we had originally projected. The extra bump in sea level came because the Gulf Stream had thermally expanded and was slowing down as a result of warmer ocean surface temperatures.[13] The new estimates for 2050, once you included all the sources of the rise in sea level—from Greenland, from Antarctica, from glaciers and ice caps, and from thermal expansion—as well as the dynamic effects, could be as high as 3 feet.[14] And adding as little as 1.5 feet of sea level rise to the storm surge of a category 3 hurricane on a worst-case storm track would devastate many parts of the city—the Rockaways, Coney Island, much of southern Brooklyn and Queens, portions of Long Island City, Astoria, Flushing Meadows–Corona Park, Queens, Lower Manhattan, and eastern Staten Island from Great Kills Harbor north to the Verrazano Bridge would be underwater. Thank God we had built those storm surge barriers and seawalls.

## August 2050

Not everyone loves New York. But those who do love it love it intensely. And through some combination of luck and high-tech ingenuity, those who loved the city ultimately saved it. In 2050, when Hurricane Xavier—a category 4 monster, which sprang up from the bathtub that the Atlantic had become finally arrived—people sat back and watched it like the World Series. We knew we had a home team advantage, just like the Yankees.

# EPILOGUE

# THE TRILLIONTH TON

**The large fields and acres produced no grain**
**The flooded fields produced no fish**
**The watered gardens produced no honey and wine**
**The heavy clouds did not rain. . . .**
**On its plains where grew fine plants**
**"lamentation reeds" now grew.**

—"The Curse of Akkad," c. 4110 BP

"The Curse of Akkad," an epic poem known as a *city lament*, is believed to have been written by a Sumerian priest after the collapse of the Akkadian empire, about 4,200 years ago. The lament is considered a work of literature, although some archaeologists believe it describes actual events that detail the fall of the world's first empire. Whatever it is, the lament is probably one reason why I became a climatologist. I wanted to know if some of those details included an abrupt change in climate.

In its heyday 4,300 years ago, the Akkadian empire, under the rule of Sargon of Akkad, stretched across Mesopotamia from the Persian Gulf to the headwaters of the Tigris and Euphrates rivers. In addition to deserving the title "world's first emperor," Sargon probably also deserves credit for the world's first merger and acquisition. He successfully merged the remote agricultural hinterlands of northern Mesopotamia with the urban Babylonian city-states such as Kish, and Ur and Uruk in the south. He acquired those the old-fashioned way, with a large army.

It was a good setup, while it lasted. Northern Mesopotamia was prized for its rain-fed agriculture and served as a critical supplier of grain to the economic hubs to the south. Tell Leilan, a provincial northern capital, was part of the breadbasket Sargon had come to depend on. It served as a central processing point, distributing grain throughout his growing empire, as well as his growing army. Today, Tell Leilan, in northeastern Syria, is the site of a small Kurdish village. Tucked away in a corner of the modern Tell Leilan rests what remains of this great ancient capital. Harvey Weiss, a professor at Yale University, returns here every few years to excavate. I had been working with Weiss on the collapse of Akkadia for more than four years before I ever saw what remains of this ancient Akkadian capital.

You could say that my experiences, until I joined Weiss and his team of archaeologists on a dig in the summer of 1999, had been only vicarious. I had studied ancient Near Eastern history, modern Middle East policy, Arabic, and of course, climate. I was what you might call book smart. I could tell you the average high temperature and wind speed for northeastern Syria in July, but that didn't mean much until I stood on a dig site for eight hours a day in 110°F heat taking in mouthfuls of windblown dust. I may have been there in my head, but the real experience was a whole different ball game. As with most things in life, including global warming, there is something important to be said for personal experience.

Weiss's previous excavations had shown that between 4,600 and 4,400 years ago, Tell Leilan grew about sixfold in size, from 37 acres to more than 200 acres.[1] The city's residential quarters showed signs of urban planning, including straight streets lined with potsherds and with drainage lanes; and its acropolis contained several storerooms for grain distribution. But sometime around 4,200 years ago, this society began to fall apart and archaeological evidence indicates a mass exodus from Tell Leilan to points south. Tell Leilan was abandoned and sat empty for some 300 years. It was probably during this abandonment period that the city lament was composed. After less than 100 years, the world's first empire was gone.

At least, this was the story when I signed on to research the Akkadian collapse. Weiss had assembled enough archaeological evidence to suggest, at least to him, that the Akkadian empire had collapsed abruptly because of a rapidly changing climate. However, he had no data on climate to support this theory. And without such data, the story of Tell Leilan was just that, a sad story told in the form of an epic poem. With the help and guidance of Peter deMenocal, one of my PhD advisers at Columbia University, I obtained the top 6 feet of a long ocean sediment core that had been pulled up from the bottom of the Gulf of Oman. We intended to use the core to reconstruct the climate of Mesopotamia over the geologic period known as the Holocene, which spans the last 10,000 years. By taking samples at half-inch intervals down the entire length of the core, I was able to build a history of Mesopotamian climate in 100-year steps, using only dried mud from the bottom of the sea. The ocean floor remembers everything.

Mesopotamia is a notoriously dusty place, and the dust there is predominantly composed of a mineral, dolomite. Mesopotamia is also a notoriously windy place. When its steady southwest wind, called the *shamal*, kicks up, that dust is transported and dumped into the Persian Gulf and the Gulf of Oman. Technically, my mud from the Gulf of Oman was actually Mesopotamian dust. Everything has to come from somewhere.

I spent my first semester of grad school in the lab at Lamont-Doherty Earth Observatory, crushing dried core samples and measuring the concentration of dolomite in each sample. The goal was to reconstruct a history of Mesopotamian drought. On the basis of soil characteristics and prevailing winds, the more dolomite in my mud sample, the more drought in Mesopotamia, and the more trouble for Sargon.

Grinding, measuring, and running about 200 samples consumed

the better part of my life for more than a year. When the work was finally complete, I had an elegant spike about one-third of the way down the core to show for my effort.[2] According to the X-ray diffractometer (XRD) analysis, this spike meant the amount of dolomite in that one particular sample was six times higher than it had been throughout the rest of the Holocene. That spike suggested that a severe drought had gripped Mesopotamia for more than a century. And according to the AMS dates, the spike in dolomite occurred very close to 4,200 years ago, at about the time a weary Sumerian priest was sitting down to write an epic poem of collapse. Weiss had taken the city lament at its word, and our ocean sediment core suggests he was right. The world's first empire stood for just 100 years. In the end, it amounted to less than 1 inch of mud at the bottom of the ocean.

––––––

The United Nations Framework Convention on Climate Change (UNFCCC), an international environmental treaty crafted at the Earth Summit, held in Rio de Janeiro in 1992, was ratified by 192 countries, including the United States.[3] The stated objective of the Convention (Article 2) is to stabilize atmospheric concentrations of greenhouse gas at a low enough level to "prevent dangerous human interference with the climate system." The words *dangerous human interference*, carefully selected by diplomats and policy makers, are not the language of scientists.

"One of the things that has always been difficult about the concept of dangerous human interference is that it involves value judgments," explains Susan Solomon, a senior scientist with NOAA. But in 1996, using the best available science, the members of the European Union decided to make a value judgment. They agreed that a temperature increase of 3.6°F above the preindustrial global average temperature constituted dangerous human interference with the climate system. And in order to avoid severe, widespread impacts they argued that there was a need to keep global warming below that level.

Solomon knows that workings of the policy world quite well. She cochaired Working Group 1 for the IPCC Fourth Assessment Report, and research done earlier in her career on the ozone hole provided the scientific foundation for the UN Montreal Protocol, the international agreement to protect the ozone layer. Solomon had helped prove the existence and cause of the ozone hole after leading an expedition to McMurdo Sound, Antarctica, in 1986. The ozone hole appears in the early spring, so Solomon and her team spent months in Antarctica, enduring brutally cold temperatures and nearly twenty-four-hour darkness, in order to observe the hole as it formed. She and her team were able to gather enough data to provide the first evidence that enhanced levels of chlorine oxide from the chlorofluorocarbons (CFCs) were the primary cause of the ozone hole.[4] The CFCs, stable compounds used in refrigerators, in air conditioners, and as a propellant in aerosol cans, were reacting with the clouds in the stratosphere to destroy the protective ozone layer. Solomon and her colleagues had the data to prove it.

Global warming is somewhat less straightforward. "How do you decide what a dangerous level of human interference means?" Solomon asks. "It's been a real challenge as far as what science can do to inform that process without becoming, frankly, unscientific," she explains. "It's not as clear-cut as observing an Antarctic ozone hole form." That's why Solomon chose to focus on another principle of the UNFCCC. Article 3 of the Convention emphasizes "threats of serious or *irreversible* damage."

"And I can tell you from the amount of time I spent around the diplomatic folks during the IPCC process that when they say 'serious *or* irreversible damage,' that's what they mean. If they meant to say serious *and* irreversible, they would have said so. They probably spent five days negotiating whether it was going to be serious 'or' or serious 'and,' " she adds. Solomon was struck by the word *irreversible*. "Irreversible is something that doesn't involve any value judgments. Something is either irreversible or it's not," she says. "That is a piece of the problem that science can inform." And in a recent research paper, Solomon looked at the extent to which human influence on the climate is irreversible. What she discovered surprised even her.

Until recently, most scientists were working under the assumption that if we went cold turkey and brought $CO_2$ emissions to zero, $CO_2$ concentrations measured in parts per million (ppm) in the atmosphere would peak and then fall most of the way down toward preindustrial levels in about 100 to 200 years, with the warming decreasing along with them. Solomon and her colleagues, using a climate model known as an Earth-system model of intermediate complexity (EMIC), wanted to see for themselves how long it would take for the concentration and climate to head down. An EMIC is not as fancy as a general circulation model, but it has the advantage of being fast. So Solomon was able to run very long simulations of the Earth's climate and see what the atmosphere remembered of us 1,000 years from now, in the year 3000. The experiment tested what would happen if $CO_2$ emissions suddenly stopped after peaking at different concentrations, ranging from 450 to 1,200 ppm.[5] In their model, $CO_2$ levels dropped so slowly that by the year 3000 the atmospheric concentration was still substantially above preindustrial levels. Global temperatures also stayed high. The atmosphere turns out to have a better memory than we thought.

"Somebody said to me recently that I've introduced two words into this debate that just weren't there before. And they are *unequivocal* and *irreversible*," Solomon says. "And that's not a bad set of words." It was Solomon, armed with her thesaurus, who introduced the now famous word *unequivocal* into the IPCC Working Group I Summary for Policymakers.[6] For scientists, the statement reads like Hemingway, concise and powerful:

Warming of the climate system is *unequivocal*, as is now evident from observations of increases in global average air and ocean temperatures, widespread melting of snow and ice and rising global average sea level.

The statement reflects the kind of pure, scientific analysis and multiple lines of evidence that Solomon is passionate about. She is someone who draws a sharp line in the sand between science and politics.[7] Her personal opinions, she has said repeatedly, have no place in the policy arena.

"When we came up with *unequivocal*, some people said it wouldn't stick. It's too complicated. We thought a lot about what word to use. And we played around with words like *incontrovertible* and *undeniable*. But we didn't want to use those, because they were too political in their tone. They are not the kind of words you would see in a scientific paper," Solomon explains. "We went with it because it's not a statement about how others are reacting to the science, it's a very internal statement about the nature of the evidence," she says. This was a word that represented science, not politics.

"The fact that you have independent measurements not just of temperature, but also of sea level rise, and retreat of Arctic sea ice, and retreat of glaciers worldwide, increases in water vapor, all of which fit what we know should be happening on a warming planet," Solomon says—"that is the reason we were able to use this word." The science, in other words, speaks for itself.

Solomon's second word is no less powerful. "Article 3 of the framework convention is a recognition that things that are irreversible deserve special attention. Because it means you can't back out of them," she explains. But aside from death and taxes, *irreversibility* is not a concept we Americans tend to embrace. "The thing that makes this tricky is that just about any other pollution problem— acid rain, smog, DDT—works in a straightforward way. When you stop emitting, the problem goes away. The thing that's really hard here is that we've recognized it's not going to work that way with global warming. We're turning the dial on the global temperature as if it were a thermostat. But it only cranks one way. We don't get to turn it back down," she says. History has not been kind to those who have failed to understand the physics of the irreversible. Just look at a symbol of irreversibility, Easter Island.

Settled by Polynesians sometime around A.D. 900, Easter Island is an isolated 66-square-mile chunk of volcanic land situated in the middle of the Pacific Ocean between Peru and Australia. The volcanic tuff was carved to create the enormous stone statues, called *maoi*, that made Easter Island famous. The Polynesians used tree trunks to transport and erect the *maoi*. In fact, trees made much of life on the remote island possible. And archaeological evidence suggests that at one time Easter Island had a diverse forest. The bark of certain trees was used to make rope or beaten into cloth; other trees were used to build canoes, or to make harpoons.

The canoes and harpoons were essential, as the common dolphin, a porpoise weighing up to 165 pounds, was the main source of meat on the island. The Polynesians hunted with harpoons from large wooden canoes, far from the shore. Palms were probably the most important trees. The trunks provided sap that could be fermented to make wine, honey, and sugar. The fronds were ideal for thatching houses, and for making baskets, mats, and boat sails. It would seem as if the only things not made of wood on Easter Island were the *maoi*, which are thought to represent high-ranking ancestors. Wood was used for so many aspects of life on the island that by 1722, when the Dutch explorer Jacob Rogeveen arrived on the island, he saw no trees more than 10 feet tall. The gradual deforestation was nearly complete. Today, Easter Island serves as one of the most extreme examples of forest destruction in all of history.

Deforestation made it impossible for the Polynesians to build the canoes that allowed them to catch porpoises. Land birds, with no trees to nest in, disappeared. Palm nuts, Malay apples, and all other wild fruits were gone. The deforestation also led to soil erosion that resulted in reduced agricultural yields. It is believed that deforestation set off a chain of events that eventually led to collapse. The population dropped sharply and the construction of the *maoi* ceased. The only things left to eat on the island were rats and people. In his book *Collapse*, Jared Diamond says:

> I have often asked myself, "What did the Easter Islander who cut down the last palm tree say while he was doing it?" Like modern loggers, did he shout "Jobs, not trees!"? Or: "Technology will solve our problems, never fear, we'll find a substitute for wood"? Or: "We don't have proof that there aren't palms somewhere else on Easter, we need more research, your proposed ban on logging is premature and driven by fear mongering"? Similar questions arise for every society that has inadvertently damaged its environment, including ours.[8]

This begs the question: with regard to global warming, what is the equivalent of cutting down the last palm tree? The answer may not be simple, but if you accept the meaning of the words *unequivocal* and *irreversible*, and you accept the implied value judgment of 3.6°F warming, you can come up with a fairly good proxy. "If you accept that, you can now calculate how many gigatons of carbon you can emit and keep warming below 3.6°F," Solomon says. Scientists can boil it down to one number. And that number is 1 trillion.

Myles Allen—a professor at the University of Oxford—and his colleagues found that if we could limit all $CO_2$ emissions from fossil fuels and changes in land use to 1 trillion tons of carbon in total, there would be a good chance that the climate would not warm more than 3.6°F above its preindustrial range.[9] A companion study, by Malte Meinshausen at the Potsdam Institute, found that the world would have to limit emissions of all greenhouse gases, not just $CO_2$, to the equivalent of 400 gigatons of carbon between 2000 and 2050 in order to stand a 75 percent chance of avoiding more than 3.6°F of warming.[10] The other greenhouse gases, such as methane and nitrous oxide, are expected to produce as much warming as 125 gigatons of carbon in the form of $CO_2$; that means emissions of $CO_2$ itself over the next forty years have to add up to less than 275 gigatons of carbon. No one said this was going to be easy.

That is especially true when you consider how much $CO_2$ we've put up there already. Since the start of the industrial revolution about 250 years ago, we've burned about half of the 1 trillion tons. Global emissions currently average about 9 billion tons a year, and they're rising. In May 2009, the Energy Information Administration (EIA) released *International Energy Outlook 2009*, an annual report that projects energy trends through 2030. The report concluded that with no further policies to reduce $CO_2$ emissions, total global emissions will reach about 11 billion tons of carbon by 2030. The jump in emissions is attributed to a projected 44 percent increase in energy demand by 2030, much of it produced from fossil fuels. Renewable energy is expected to grow fastest, but fossil fuels will continue to serve as the dominant source of energy to meet the growing demand coming from developing nations. The report says 94 percent of the increase in industrial energy use between now and 2030 is expected to take place in developing countries; Brazil, Russia, India, and China are expected to account for two-thirds of that growth.[11]

The implications are simple: the more $CO_2$ we dump into the atmosphere, the warmer it gets, and the more serious *and* irreversible the damage. If carbon emissions were trees, then the more we've cut down by 2020, the fewer will be left to cut by 2050. In essence, the 1 trillion-ton limit allows the world to follow its current trend for about forty more years before having to quit carbon cold turkey, all at once. Somewhere out there, in a coal seam in Wyoming or an oil field in Saudi Arabia, sits the trillionth ton. Unless scientists come up with some way to suck $CO_2$ out of the atmosphere or a cheap form of renewable energy, the fate of that trillionth ton rests in the hands of policy makers, not scientists.

In December 2009, seventeen years after the Earth Summit in

Rio de Janeiro, policy makers and scientists gathered once again at the Conference of the Parties (COP 15) in Copenhagen, Denmark. Since the ratification of the UNFCCC, they have been using these conferences to assess progress and establish legally binding obligations for countries to reduce their greenhouse gas emissions. It was at COP 3 in 1997 that the Kyoto Protocol first set binding targets for reducing greenhouse gas emissions. The overall goal for COP 15 was to establish a new global climate agreement to replace the Kyoto Protocol when it expires in 2012. Dozens of countries, including China and the United States—the top two carbon polluters—came to Copenhagen with proposals to cut their emissions. But in the end, world leaders left the meeting with a deal that was seen as weak and lacking in detail. Although it did set up the first major program of aid to help poorer nations adapt to climate change, it offered few specifics in terms of pollution reductions. Upon the announcement of the deal, a team of experts led by a professor at MIT made a quick calculation that converted the language of policy makers into the language of scientists. As it stood, the deal was likely to result in a 5.7°F rise in average global temperature. In the end, the words of science—*unequivocal* and *irreversible*—were still not powerful enough to shift the forecast for the future.

# APPENDIX 1

# UNITED STATES CLIMATE CHANGE ALMANAC

Clearly, climate change will leave no part of the world untouched; but just as all the places I wrote about have their own indicators showing that climate change is taking place, so does your town or city. What follows here is an almanac that includes current and future temperature trends for a variety of cities around the United States. I'll start with a look at the big picture in the United States.

## 1. Total Monthly Record High versus Low Temperatures for the United States

One way to get an idea of how climate change is making itself felt in terms of day-to-day weather in the United States is to look at the total number of daily record high and low temperatures that have been set around the country. A *record daily high* means that the high temperature recorded at a specific weather station was higher on a specific day than on that same day in previous years. A *record daily low* means that the lowest temperature on a specific day at a specific weather station was lower than on that same day in previous years.

In a recent study, scientists gathered and tallied records from across the United States to get a better picture of the long-term changes in record highs and record lows.[1] They analyzed millions of daily high and low temperature readings taken over a period of six decades at about 1,800 weather stations across the country. The temperature measurements were collected by the National Climatic Data Center of the National Oceanic and Atmospheric Administration (NOAA) and underwent a strict quality control process that can pick out potential problems, such as missing data and inconsistent readings caused by changes in thermometers and station locations. What makes the tables below quite fascinating is that there is such a large discrepancy between the total number of record high temperatures and the total number of record low temperatures. Technically, if the Earth's temperature was not increasing, you would expect that the number of record daily highs and lows being set each year should be about even. But that is far from the case. For the period from January 1, 2000, to December 20, 2009, the continental United States set 294,276 record highs and only 145,498 record lows. And if you look back over the past sixty years, that picture is reinforced. The ratio of record daily high to record daily low temperatures was almost one to one in the 1950s but has been rising steadily since the 1980s.

As you'll notice in the tables below, there is still a lot of variability from year to year. In fact, October 2009 serves as an excellent example of a month that was quite cold: the number of record lows was 1,908 while the number of record highs was only 1,017. On average, October 2009 was the third-coolest October since record keeping began in 1895. The average October temperature of 50.8°F was 4.0°F below the twentieth-century average. But when you step back and look at the big picture, global land and ocean surface temperature for October 2009 was the sixth-warmest on record, with an anomaly of 1.03°F above the twentieth-century average of 57.1°F. The lesson here is that there will always be natural variability in space and time!

What's also interesting is that the current two-to-one ratio of highs versus lows across the country comes from the fact that there is a relatively small number of record lows. In other words, much of the nation's warming is taking place at night, and, as a result, temperatures dip down less often to set new record lows. This finding is completely consistent with climate models showing that higher overnight lows are to be expected as the planet warms.

In addition to looking at historical temperatures in recent decades, scientists also used a computer model to simulate how record high and low temperatures are likely to change in the future. The model indicates that if greenhouse gas emissions continue to increase in a "business-as-usual" scenario, the U.S. ratio of daily record high to record low temperatures would increase to about twenty to one by 2050 and fifty to one by 2100.

## Number of Record High and Low Temperatures for the United States for 2000

| Month | 2000 High | 2000 Low |
|---|---|---|
| January | 4,480 | 110 |
| February | 3,758 | 402 |
| March | 4,259 | 189 |
| April | 2,082 | 934 |
| May | 6,287 | 887 |
| June | 2,222 | 1,681 |
| July | 2,953 | 1,949 |
| August | 4,295 | 979 |
| September | 6,306 | 2,813 |
| October | 2,123 | 4,251 |
| November | 689 | 2,810 |
| December | 601 | 1,866 |
| Total | 40,055 | 18,871 |

# Number of Record High and Low Temperatures for the United States for 2001

| Month | 2001 High | 2001 Low |
|---|---|---|
| January | 1,299 | 1,152 |
| February | 1,339 | 466 |
| March | 732 | 877 |
| April | 3,202 | 1,942 |
| May | 4,830 | 1,713 |
| June | 1,633 | 1,356 |
| July | 2,038 | 1,797 |
| August | 3,699 | 503 |
| September | 2,246 | 1,586 |
| October | 2,109 | 2,262 |
| November | 4,001 | 339 |
| December | 1,724 | 241 |
| Total | 28,852 | 14,234 |

# Number of Record High and Low Temperatures for the United States for 2002

| Month | 2002 High | 2002 Low |
|---|---|---|
| January | 4,814 | 495 |
| February | 2,033 | 1,261 |
| March | 1,208 | 3,956 |
| April | 5,518 | 1,530 |
| May | 2,242 | 6,487 |
| June | 2,687 | 935 |
| July | 4,079 | 623 |
| August | 3,833 | 2,299 |
| September | 2,176 | 358 |
| October | 1,142 | 2,314 |
| November | 2,144 | 1,514 |
| December | 1,817 | 728 |
| Total | 33,693 | 22,500 |

# Number of Record High and Low Temperatures for the United States for 2003

| Month | 2003 High | 2003 Low |
|---|---|---|
| January | 3,969 | 917 |
| February | 1,176 | 1,227 |
| March | 2,113 | 1,276 |
| April | 2,530 | 948 |
| May | 2,896 | 1,299 |
| June | 1,157 | 1,409 |
| July | 4,852 | 923 |
| August | 3,630 | 326 |
| September | 1,677 | 657 |
| October | 7,699 | 966 |
| November | 2,925 | 1,777 |
| December | 1,080 | 260 |
| Total | 35,704 | 11,985 |

# Number of Record High and Low Temperatures for the United States for 2004

| Month | 2004 High | 2004 Low |
|---|---|---|
| January | 1,475 | 1,124 |
| February | 547 | 834 |
| March | 6,242 | 365 |
| April | 1,958 | 623 |
| May | 2,524 | 1,275 |
| June | 1,253 | 1,752 |
| July | 742 | 2,509 |
| August | 724 | 5,186 |
| September | 1,122 | 1,025 |
| October | 1,514 | 517 |
| November | 796 | 589 |
| December | 1,801 | 1,281 |
| Total | 20,698 | 17,080 |

# Number of Record High and Low Temperatures for the United States for 2005

| Month | 2005 High | 2005 Low |
|---|---|---|
| January | 3,805 | 743 |
| February | 2,539 | 158 |
| March | 2,266 | 599 |
| April | 1,341 | 1,727 |
| May | 2,627 | 2,290 |
| June | 1,556 | 836 |
| July | 3,714 | 1,354 |
| August | 2,286 | 649 |
| September | 4,341 | 642 |
| October | 2,426 | 898 |
| November | 4,595 | 826 |
| December | 2,553 | 1,986 |
| Total | 34,049 | 12,708 |

# Number of Record High and Low Temperatures for the United States for 2006

| Month | 2006 High | 2006 Low |
|---|---|---|
| January | 3,805 | 743 |
| February | 2,539 | 158 |
| March | 2,266 | 599 |
| April | 1,341 | 1,727 |
| May | 2,627 | 2,290 |
| June | 1,556 | 836 |
| July | 3,714 | 1,354 |
| August | 2,286 | 649 |
| September | 4,341 | 642 |
| October | 2,426 | 898 |
| November | 4,595 | 826 |
| December | 2,553 | 1,986 |
| Total | 34,049 | 12,708 |

# Number of Record High and Low Temperatures for the United States for 2007

| Month | 2007 High | 2007 Low |
|---|---|---|
| January | 2,279 | 1,531 |
| February | 872 | 1,779 |
| March | 6,956 | 933 |
| April | 1,556 | 3,145 |
| May | 2,494 | 889 |
| June | 1,852 | 988 |
| July | 3,165 | 1,429 |
| August | 6,641 | 519 |
| September | 3,014 | 1,237 |
| October | 4,145 | 549 |
| November | 1,570 | 267 |
| December | 1,832 | 390 |
| Total | 36,376 | 13,656 |

# Number of Record High and Low Temperatures for the United States for 2008

| Month | 2008 High | 2008 Low |
|---|---|---|
| January | 1,813 | 528 |
| February | 1,399 | 533 |
| March | 490 | 540 |
| April | 389 | 1,734 |
| May | 1,281 | 1,036 |
| June | 1,585 | 534 |
| July | 494 | 658 |
| August | 949 | 519 |
| September | 507 | 555 |
| October | 733 | 1,099 |
| November | 1,564 | 883 |
| December | 1,654 | 556 |
| Total | 12,858 | 9,175 |

# Number of Record High and Low Temperatures for the United States for 2009

| Month | 2009 High | 2009 Low |
|---|---|---|
| January | 1,292 | 642 |
| February | 1,556 | 392 |
| March | 1,696 | 855 |
| April | 1,602 | 746 |
| May | 1,074 | 847 |
| June | 1,167 | 538 |
| July | 1,274 | 1,904 |
| August | 1,144 | 1,205 |
| September | 1,028 | 360 |
| October | 1,017 | 1,908 |
| November | 1,096 | 143 |
| December | 264 | 1,086 |
| Total | 14,210 | 10,626 |

# Total Number of Record High and Low Temperatures for the United States from 2000 to 2009

| Year | High | Low |
|---|---|---|
| 2000 | 40,055 | 18,871 |
| 2001 | 28,852 | 14,234 |
| 2002 | 33,693 | 22,500 |
| 2003 | 35,704 | 11,985 |
| 2004 | 20,698 | 17,080 |
| 2005 | 34,049 | 12,708 |
| 2006 | 37,465 | 14,762 |
| 2007 | 36,376 | 13,656 |
| 2008 | 12,858 | 9,175 |
| 2009 | 14,210 | 10,626 |
| Total | 293,960 | 145,597 |

## 2. Increasing Number of Hot Days in U.S. Cities

The bar plots that follow illustrate how extremely hot days are likely to become more common and more intense as the overall climate warms during the next century. They also help demonstrate that a general warming of the climate has significant implications for the number and severity of extreme weather and climate events: in this case, heat waves.

For selected large cities or regions in the United States, the plots compare the average number of extremely hot days observed during the summer months in the twentieth century with projections for extremely hot days in the middle and end of the twenty-first century. Climate research indicates that heat waves are likely to be more stifling, and potentially more deadly, in coming decades as climate change progresses. For many locations, heat waves are projected to be more frequent, more intense, and longer lasting, and extreme heat events that are currently considered rare will become more common in coming years.[2]

Extreme weather and climate events can cause significant damage, and heat waves are considered public health emergencies because of their effects on human health. Hot temperatures contribute to increased emergency room visits and hospital admissions for cardiovascular disease and can cause heatstroke and other life-threatening conditions. The elderly are particularly vulnerable to extreme heat.

Heat waves such as the Chicago heat wave of 1995 and the European heat wave of 2003, which killed an estimated 50,000 people, have proved especially deadly to vulnerable populations, including the elderly and persons with respiratory illnesses.[3]

Scientists at Climate Central conducted their own special analysis to generate the values in these graphics, using techniques and general climate projections that are well established in the peer-reviewed scientific literature.[4]

The projections for extreme heat in the years 2050 and 2090

are based on an average of twelve computer models that simulate climate. For these projections, Climate Central used a scenario of moderate-high greenhouse gas emissions, which currently appears optimistic, since global emissions have exceeded this scenario in recent years. Climate Central used a common technique to translate large-scale climate information from the computer models to provide useful information about local and regional conditions. This method involves calculating differences between time series data from current and future global climate model simulations and then adding these changes to time series of observed climate data.

Scientists at Climate Central first identified weather observation stations closest to each city, as well as the closest point in the output of computer models, which is known as a *grid point*. For the station data, Climate Central examined temperature information for the summer months (June, July, and August) during two twenty-year periods to determine how extreme heat events have evolved during the twentieth century. Those periods were 1951–1970 and 1981–2000.

For the computer model data, Climate Central looked at two future twenty-year periods of projected maximum temperature data, 2046–2064 and 2081–2100, as well as the simulated current climate period 1981–2000. These data shed light on how extreme heat events may evolve as the climate changes. The analyses are based on recent data from weather stations, regional-scale outputs from climate projection models, and a common technique for deducing best-guess local climate projections from regional projections. This method involves calculating differences between current and future global climate model simulations, and applying them to observed climate data from the same vicinity.

For a given month, Climate Central calculated changes in the twenty-year average monthly maximum temperature between the two periods in the station data and between future time periods in the computer simulations and the current climate simulation. This provided a comparison of the twenty-year climatology at the end of the twentieth century and the earlier period 1951–1970. It also permitted a comparison of the model-simulated average monthly maximum temperature in 2046–2064 with that of 1981–2000, and the end-of-the-century period 2081–2100 with 1981–2000.

Because data from twelve different computer models were used, Climate Central took a model average of the differences in simulated temperature changes between the time periods. Next, Climate Central turned to the station observations and used the daily temperature data for the period 1981–2000 to determine a climatology of daily temperature values. From this climatology came the set of numbers to go into the second bar of each group of four bars. Each group pertains to a given temperature threshold of 90°F, 95°F, and 100°F. Climate Central counted how many days exceeded the temperature threshold during each of the twenty months in the climatology and then averaged those numbers.

Climate Central generated the other bars similarly, simply shifting the climatology of daily maximum temperatures for the period 1981–2000 by the temperature changes computed from the model simulations (or the 1951–1970 observed data) and repeating the count on the new sets of daily maximum temperatures that were generated.

This last step created new, simulated data for each city for twenty Augusts in the middle of the twenty-first century. The same method that was used with actual 1981–2000 temperatures to estimate the average number of days over each temperature threshold in this future scenario was then applied.

The resulting projections give long-term averages, not predictions for any individual year; actual outcomes will vary significantly from year to year, owing to the natural variability of climate. Furthermore, because the modeling and methods used involve uncertainty, the projections should be taken as best guesses within a range of uncertainty. True long-term averages will be likely to prove somewhat higher or lower than the projections here. However, all twelve models are unanimous in projecting increased hot days (relative to the present) by the middle of the twenty-first century.

All model outputs used were based on a scenario of medium-high greenhouse gas emissions, called A1B by the Intergovernmental Panel on Climate Change. Carbon dioxide and other greenhouse gas emissions during the decade spanning 2000–2010 have already exceeded the A1B scenario, so the projections here are conservative and represent a future in which greenhouse gas emissions are reduced compared with the current trend.

The number of hot July and August days (above 90°F, 95°F, and 100°F) in twenty U.S. cities over four time periods (1951–1970, 1981–2000, 2046–2064, 2081–2100) are presented below. Cities are listed in order of population size. For a given month (in this case, July or August), changes in the twenty-year average monthly number of high temperature are represented by four columns. The first two columns represent two distinct twenty-year periods (1951–1970, labeled 1960; and 1981–2000, labeled 1990) in the station data. The second two columns represent best guesses about the future, with column three representing the twenty-year climatology in 2046–2064 (labeled 2050) compared with the twenty-year climatology in 1981–2000, both model derived; and column four representing the twenty-year climatology in 2081–2100 (labeled 2090) compared with the twenty-year climatology in 1981–2000, both model derived.

# New York

**Number of hot July days in New York**

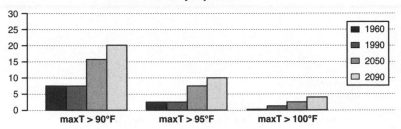

**Number of hot August days in New York**

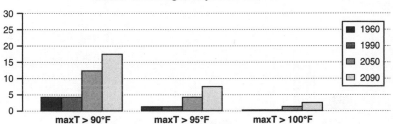

# Los Angeles

**Number of hot July days in Los Angeles**

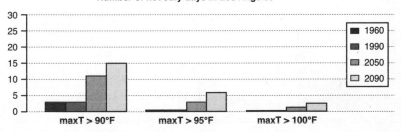

**Number of hot August days in Los Angeles**

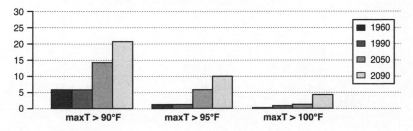

# Chicago

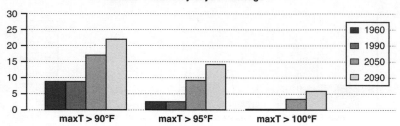

# Philadelphia

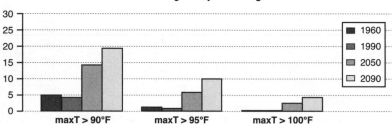

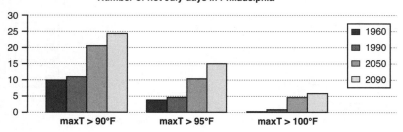

# Dallas

### Number of hot July days in Dallas

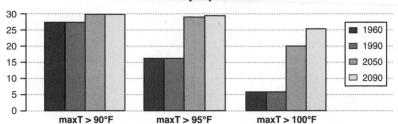

### Number of hot August days in Dallas

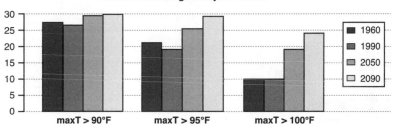

# San Francisco

### Number of hot July days in San Francisco

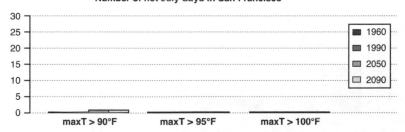

### Number of hot August days in San Francisco

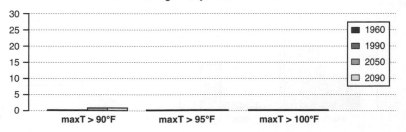

# Boston

### Number of hot July days in Boston

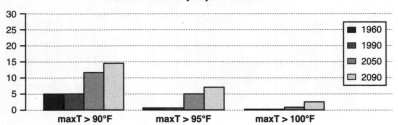

### Number of hot August days in Boston

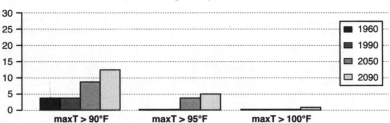

# Atlanta

### Number of hot July days in Atlanta

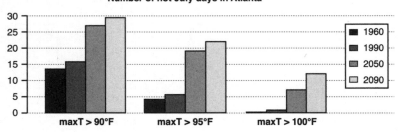

### Number of hot August days in Atlanta

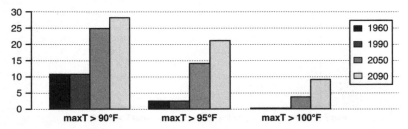

# Washington, D.C.

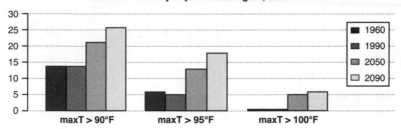

Number of hot July days in Washington, D.C.

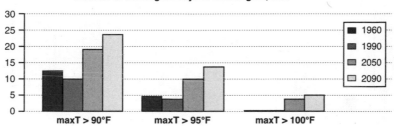

Number of hot August days in Washington, D.C.

# Houston

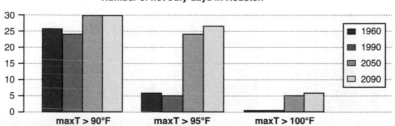

Number of hot July days in Houston

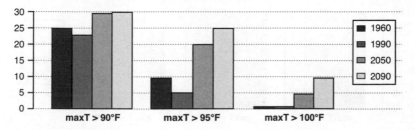

Number of hot August days in Houston

# Detroit

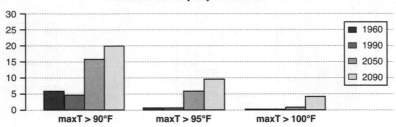

Number of hot July days in Detroit

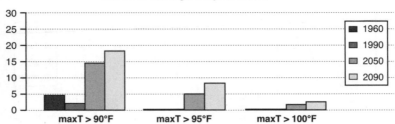

Number of hot August days in Detroit

# Phoenix

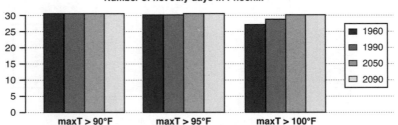

Number of hot July days in Phoenix

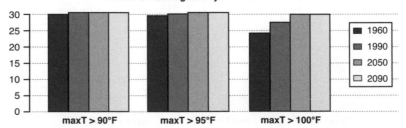

Number of hot August days in Phoenix

# Tampa

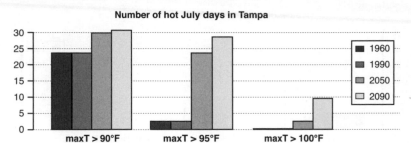

**Number of hot July days in Tampa**

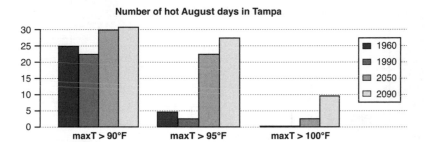

**Number of hot August days in Tampa**

# Seattle

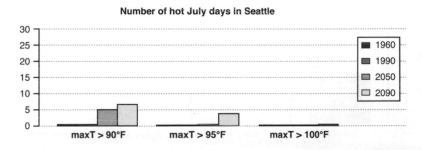

**Number of hot July days in Seattle**

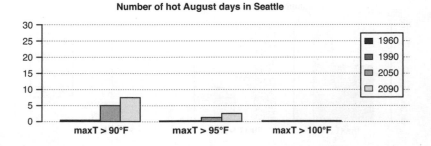

**Number of hot August days in Seattle**

# Minneapolis

### Number of hot July days in Minneapolis

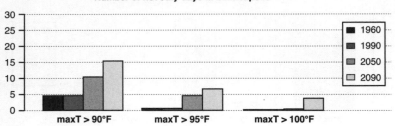

### Number of hot August days in Minneapolis

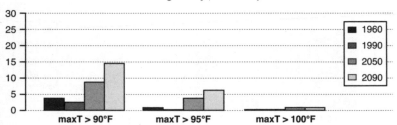

# Miami

### Number of hot July days in Miami

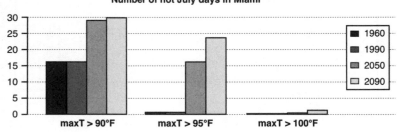

### Number of hot August days in Miami

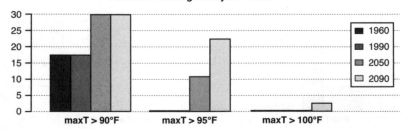

# Cleveland

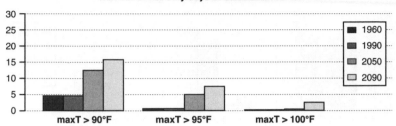

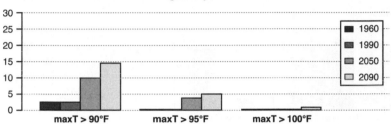

# Denver

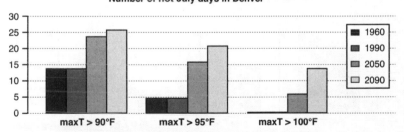

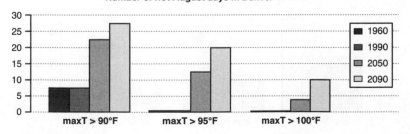

# Orlando

### Number of hot July days in Orlando

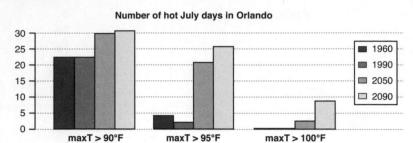

### Number of hot August days in Orlando

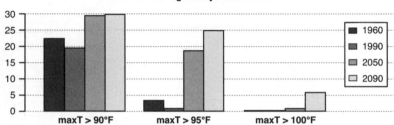

# Sacramento

### Number of hot July days in Sacramento

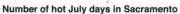
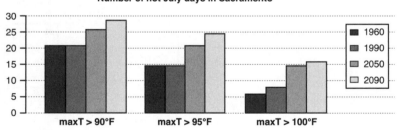

### Number of hot August days in Sacramento

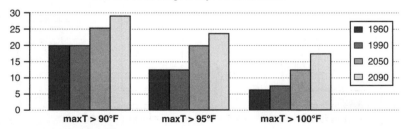

# 3. Temperature Trends by State

It's always useful to look at the long-term temperature trend at a more regional level. For example, the table below shows how much each state in the continental United States has warmed, on average, over the period 1976–2005. The first column shows the long-term trend averaged over the entire year (January to December). The second column shows the long-term trend for *winter* (December to February). You'll notice that winters have warmed significantly.

| 1976–2005 | Annual | Winter | 1976–2005 | Annual | Winter |
|---|---|---|---|---|---|
| Alabama | 1.41°F* | 3.81°F* | Nevada | 1.58°F* | 0.4°F |
| Arizona | 1.97°F* | 0.62°F | New Hampshire | 1.76°F* | 3.22°F* |
| Arkansas | 1.06°F* | 3.75°F* | New Jersey | 1.85°F* | 4.13°F* |
| California | 1.56°F* | 0.27°F | New Mexico | 1.73°F* | 1.59°F |
| Colorado | 1.64°F* | 2.31°F | New York | 2.18°F* | 3.94°F* |
| Connecticut | 1.57°F* | 3.58°F* | North Carolina | 1.32°F* | 3.28°F* |
| Delaware | 1.77°F* | 4.04°F* | North Dakota | 0.89°F | 4.58°F |
| Florida | 1.21°F* | 2.79°F* | Ohio | 1.89°F* | 5.2°F* |
| Georgia | 0.76°F | 2.82°F* | Oklahoma | 0.58°F | 3.4°F* |
| Idaho | 2.12°F* | 2.74°F | Oregon | 1.42°F* | 1.45°F |
| Illinois | 2.01°F* | 6.26°F* | Pennsylvania | 2.14°F* | 4.48°F* |
| Indiana | 1.96°F* | 5.86°F* | Rhode Island | 1.96°F* | 4.24°F* |
| Iowa | 0.93°F | 4.8°F* | South Carolina | 1.07°F* | 2.86°F* |
| Kansas | 0.61°F | 3.47°F | South Dakota | 1.11°F | 5.25°F |
| Kentucky | 2.06°F* | 5.47°F* | Tennessee | 1.81°F* | 4.8°F* |
| Louisiana | 1.65°F* | 3.79°F* | Texas | 1.32°F* | 3.09°F* |
| Maine | 0.64°F | 1.75°F | Utah | 2.14°F* | 2.31°F |
| Maryland | 1.61°F* | 3.92°F* | Vermont | 2.37°F* | 3.74°F* |
| Massachusetts | 1.58°F* | 3.56°F* | Virginia | 1.04°F* | 3.28°F* |
| Michigan | 1.91°F * | 4.75°F * | Washington | 2.23°F* | 3.06°F* |
| Minnesota | 1.48°F | 5.7°F* | Washington, D.C. | 1.11°F* | 3.31°F* |
| Mississippi | 1.3°F* | 3.59°F* | West Virginia | 1.65°F* | 4.15°F* |
| Missouri | 1.2°F | 5.03°F* | Wisconsin | 1.96°F* | 6.01°F* |
| Montana | 1.54°F* | 3.97°F | Wyoming | 1.52°F* | 2.79°F |
| Nebraska | 1.45°F | 4.97°F* | | | |

*Asterisk denotes statistical significance at the 90 percent confidence level. (SOURCE: C. TEBALDI, CLIMATE CENTRAL)

# APPENDIX 2

# NEW YORK STATISTICS

## New York City Area: Hurricanes

| Date | Name | Category* |
|---|---|---|
| Sept. 3–5, 1815 | Great September Gale of 1815 | 3 |
| Sept. 3, 1821 | unnamed | 1–2 |
| Sept. 1869 | New England Storm | 1 |
| Aug. 23, 1893 | Eastern New England Storm | 1 |
| Aug. 23, 1893 | Midnight Storm | 1–2 |
| Sept. 21, 1938 | Long Island Express/New England Storm | 3 |
| Sept. 15, 1944 | unnamed | 1 |
| Aug. 1954 | Carol | 3 |
| Sept. 12, 1960 | Donna | 3 |
| Sept. 21, 1961 | Esther | 1–2 |
| June 1972 | Agnes | 1 |
| Aug. 10, 1976 | Belle | 1 |
| Sept. 27, 1985 | Gloria | 2–3 |
| Aug. 1991 | Bob | 2 |
| Sept. 1999 | Floyd | 2 |

*Category column represents the estimated storm maximum; not necessarily experienced in the region of New York City. (SOURCE: NPCC, CLIMATE RISK INFORMATION)

# New York City Transportation Facilities Potentially Most Vulnerable to Inundation from Climate Change

| Facility | Elevation (feet) |
|---|---|
| **Metropolitan Transit Authority (MTA)** | |
| Christopher Street Station (1 line) | −14.6 |
| Canal Street Station (A, C, E lines) | 8.7 |
| South Ferry Station (1 line) | 9.1 |
| Long Island Railroad Far Rockaway Station | 9.2 |
| Verrazano Narrows Bridge | 8 |
| MetroNorth Hudson Line tracks, Croton River | 7.0–7.5 |
| **Port Authority of New York and New Jersey** | |
| Holland Tunnel NY Entrance | 9.5 |
| LaGuardia Airport | 6.8 |
| Port Newark and Elizabeth | 9.6 |
| **NYC Department of Transportation** | |
| Battery Park Tunnel | 9 |
| FDR Drive above 59th Street | 6 |

(SOURCE: R. ZIMMERMAN[5])

# THE WORLD'S MOST VULNERABLE PLACES

On the basis of my discussions with dozens of climate experts, I selected just a few vulnerable places from around the world to showcase the specific regional risks associated with climate change. There are many vulnerable places that I was, of course, unable to discuss. But Mike MacCracken, the chief scientist for Climate Change Programs at the Climate Institute, has assembled an excellent list of the top ten prevailing threats associated with climate change as well as examples of the places that are most vulnerable to these threats.[6] The threats are listed, in no specific order, below.

### 1. Sea, Salt, and Storms

*River deltas, bays, and estuaries:* Dhaka, Cairo, New Orleans, Sacramento–San Joaquin Delta, Hong Kong, Chesapeake Bay, New York, London.

*Low-lying coastal plains:* Miami, Charleston, Boston, Long Island, Amsterdam, Venice, Rotterdam, Venice.

*Barrier islands:* Alaskan villages, North Carolina coast.

*Island nations:* Fiji, Tahiti, Tuvalu.

## 2. Acidification
*Coral reefs and atolls:* Maldives; Great Barrier Reef; American trust territories; Key West, Florida.

## 3. Water
*Mediterranean environments:* Sacramento, San Diego; Los Angeles; Atlanta; Las Vegas; Phoenix; Albuquerque; Dakar; Lima, Peru; Quito, Ecuador; La Paz, Bolivia; Sana'a, Yemen.

## 4. Snowmelt and Runoff
*Snow-fed rivers amid forested mountains:* Many large cities of India, Pakistan, and China; Portland and the Pacific Northwest; Sacramento–San Joaquin river basin; downstream from the Alps.

## 5. Fire and Beetles
*Arid regions:* Western United States, Canada (Alberta, British Columbia); Spain; Portugal.

## 6. Food and Mass Migration
*Agricultural plains:* United States Great Plains; Australian coastal regions; Tijuana, Mexico; Lagos, Africa; Nairobi, Kenya.

## 7. Permafrost Thaw
*Seasonal freezing:* Fairbanks, Alaska; northern Canada in general; Siberia.

## 8. Heat
*Hot and humid (summer) weather regimes:* Texas generally, New York, Chicago, Paris, southern China.

## 9. Hurricanes and Typhoons

*Tropical cyclone paths:* New Orleans, New York, Miami, Charleston, Chesapeake Bay, Hong Kong, Tokyo and other cities in Japan, Shanghai, Manila.

## 10. Dead Zones, Pollution, and Disease

Rio de la Plata/Buenos Aires: hypoxia due to upstream changes in land use and increased severe precipitation events. Mexico City: air pollution.

# NOTES

## 1. CLIMATE AND WEATHER TOGETHER

1. Kohut, A., Keeter, S., Doherty, C., Dimock, M., and Remez, M. Economy, Jobs Trump All Other Policy Priorities (Pew Research Center for the People and the Press, Washington, DC, 2009).
2. Weber, E. U., The Utility of Measuring and Modeling Perceived Risk, in *Choice, Decision, and Measurement: Essays in Honor of R. Duncan Luce*, edited by A. A. J. Marely (Lawrence Erlbaum Associates, Mahwah, NJ, 1997), pp. 45–57.
3. Weber, E., Experience-Based and Description-Based Perceptions of Long-Term Risk: Why Global Warming Does Not Scare Us (Yet). *Climatic Change* 77 (1–2), 103–120 (2006). See also Marx, S. M. et al., Communication and Mental Processes: Experiential and Analytic Processing of Uncertain Climate Information. *Global Environmental Change* 17 (1), 47–58 (2007).
4. Hirshleifer, D., and Shumway, T., Good Day Sunshine: Stock Returns and the Weather. *Journal of Finance* 58 (3), 1009–1032 (2003).
5. Sunstein, C. R., The Availability Heuristic, Intuitive Cost-Benefit Analysis, and Climate Change. *Climatic Change* 77 (1–2), 195–210 (2006).

## 2. SEEING CLIMATE CHANGE IN OUR PAST

1. Imbrie, J., and Imbrie, K. P., *Ice Ages: Solving the Mystery* (Enslow Publishers, Short Hills, NJ, 1979).
2. Weart, S. R. ed., *The Discovery of Global Warming* (Harvard University Press, Cambridge, MA, 2003).
3. Agassiz, L., *Etudes sur les glaciers* (privately published, Neuchâtel, 1840).
4. Fourier, J., Remarques générales sur les températures du Globe Terrestre et des espaces planétaires. *Annales de Chemie et de Physique* 27, 136–167 (1824).
5. Burchfield, J. A., John Tyndall—A Biographical Sketch, in *John Tyndall, Essays on a Natural Philosopher* (Royal Dublin Society, Dublin, 1981).
6. Arrhenius, S., On the Influence of Carbonic Acid in the Air upon the Temperature of the Ground. *Philosophical Magazine and Journal of Science* 41, 237–275 (1896).
7. Ruddiman, W. F., *Earth's Climate: Past and Future* (W. H. Freeman and Company, New York, 2001).

## 3. THE SCIENCE OF PREDICTION

1. Lynch, P., *The Emergence of Numerical Weather Prediction: Richardson's Dream* (Cambridge University Press, New York, 2006).

2. Somerville, R. C. J., *The Forgiving Air: Understanding Environmental Change,* 2nd ed. (American Meteorological Society, Boston, 2008).
3. Hegerl, G. C., and Zwiers, F. W., Understanding and Attributing Climate Change, in *Climate Change 2007: The Physical Science Basis. Contribution of Working Group I to the Fourth Assessment Report of the Intergovernmental Panel on Climate Change,* edited by S. Solomon et al. (Cambridge University Press, Cambridge, U.K., 2007). See also Stouffer, R. J., Hegerl, G. C., and Tett, S. F. B., A Comparison of Surface Air Temperature Variability in Three 1000-Year Coupled Ocean-Atmosphere Model Integrations. *Journal of Climate* 13, 513–537 (1999).
4. Hansen, J., et al., A Pinatubo Climate Modeling Investigation, in *The Mount Pinatubo Eruption: Effects on the Atmosphere and Climate,* edited by G. Fiocco, D. Fua, and G. Visconti (Springer-Verlag, Heidelberg, Germany, 1996), pp. 233–272.
5. Duffy, P. B., Santer, B. D., and Wigley, T. M. L., Solar Variability Does Not Explain Late-20th-Century Warming. *Physics Today* (January 2009), 48–49 (2009). See also North, G. R., Wu, Q., and Stevens, M., in *Solar Variability and Its Effect on Climate,* edited by J. M. Pap and P. Fox (American Geophysical Union, Washington, DC, 2004), Vol. Geophysical Monograph 141.
6. Thompson, D. W. J., and Solomon, S., Recent Stratospheric Climate Trends as Evidenced in Radiosonde Data: Global Structure and Tropospheric Linkages. *Journal of Climate* 18, 4785–4795 (2005). See also Ramaswamy, V. et al., Anthropogenic and Natural Influences in the Evolution of Lower Stratospheric Cooling. *Science* 311 (5764), 1138–1141 (2006).

**4. EXTREME WEATHER AUTOPSIES AND THE FORTY-YEAR FORECAST**

1. Karl, T. R., et al., *Weather and Climate Extremes in a Changing Climate: Regions of Focus* (U.S. Climate Change Science Program, 2008).
2. Schär, C., et al., The Role of Increasing Temperature Variability for European Summer Heat Waves. *Nature* 427 (6972), 332–336 (2004).
3. Stott, P. A., Stone, D. A., and Allen, M. R., Human Contribution to the European Heat Wave of 2003. *Nature* 432 (7017), 610–614 (2004).
4. Karl, T. R., Melillo, J., and Peterson, T. C., *Global Climate Change Impacts in the United States* (U.S. Climate Change Science Program, 2009).
5. Meehl, G. A., and Tebaldi, C., More Intense, More Frequent, and Longer Lasting Heat Waves in the 21st Century. *Science* 305 (5686), 994–997 (2004).

**5. THE SAHEL, AFRICA**

1. deMenocal, P. B., et al., Abrupt Onset and Termination of the African Humid Period: Rapid Climate Responses to Gradual Insolation Forcing. *Quaternary Science Reviews* 19, 347–361 (2000). See also deMenocal, P. B., Ortiz, J., Guilerson, T., and Sarnthein, M., Coherent High—and Low-Latitude Climate Variability During the Holocene Warm Period. *Science* 288, 2198–2202 (2000).
2. Shanahan, T. M., et al., Atlantic Forcing of Persistent Drought in West Africa. *Science* 324, 377–380 (2009).
3. Bobe, R., and Behrensmeyer, A. K., The Expansion of Grassland Ecosystems in Africa in Relation to Mammalian Evolution and the Origin of the Genus Homo.

*Palaeogeography, Palaeoclimatology, Palaeoecology* 207, 399–420 (2004). See also de-Menocal, P. B., Plio-Pleistocene African Climate. *Science* 270 (5233), 53–59 (1995).

4. Feakins, S. J., deMenocal, P. B., and Eglinton, T. I., Biomarker Records of Late Neogene Changes in Northeast African Vegetation. *Geology* 33 (12), 977–980 (2005).

5. deMenocal, P. B., African Climate Change and Faunal Evolution During the Pliocene-Pleistocene. *Earth and Planetary Science Letters* 220, 3–24 (2004).

6. Kandji, S. T., Verchot, L., and Mackensen, J., 2006.

7. Held, I., Delworth, T. L., Lu, J., Findell, K. L., and Knutson, T. R., Simulation of Sahel Drought in the 20th and 21st Centuries. *PNAS* 102 (50), 17891–17896 (2005).

8. Hulme, M., Climatic Perspectives on Sahelian Dessication: 1973–1998. *Global Environmental Change* 11 (19–29) (2001).

9. Herrmann, S. M., and Hutchinson, C. F., The Changing Contexts of the Desertification Debate. *Journal of Arid Environments* 63, 538–555 (2005).

10. Olsson, L., and Hall-Beyer, M., Greening of the Sahel in *Encyclopedia of Earth*, edited by C. J. Cleveland (Environmental Information Coalition, National Council for Science and the Environment, Washington, DC, 2008).

11. Hayward, D. F., and Oguntoyinbo, J. S., *The Climatology of West Africa* (Barnes & Noble, 1987).

12. Nicholson, S. E., Climatic Variations in the Sahel and Other African Regions During the Past Five Centuries. *Journal of Arid Environments* 1 (3–24) (1978).

13. Xue, Y., and Shukla, K., The Influence of Land Surface Properties on Sahel Climate. Part I: Desertification. *Journal of Climate* 6, 2232–2245 (1993).

14. Folland, C. K., Palmer, T. N., and Parker, D. E., Sahel Rainfall and Worldwide Sea Temperatures, 1901–85. *Nature* 302, 602–607 (1986).

15. Giannini, A., Saravanan, R., and Chang, P., Oceanic Forcing of Sahel Rainfall on Interannual to Interdecadal Time Scales. *Science* 302 (1027–1030) (2003).

16. IPCC, 2007. See also Patricola, C. M., and Cook, K. H., Northern African Climate at the End of the Twenty-First Century: An Integrated Application of Regional and Global Climate Models. *Climate Dynamics* (2009).

17. Biasutti, M., and Sobel, A. H., Delayed Seasonal Cycle and African Monsoon in a Warmer Climate. *Geophys. Res. Lett* submitted (2009).

18. Paeth, H., and Thamm, H. P., Regional Modelling of Future African Climate North of 15 Degrees S Including Greenhouse Warming and Land Degradation. *Climatic Change* 83 (3), 401–427 (2007).

19. Polgreen, L. In Niger, Trees and Crops Turn Back the Desert, in *The New York Times* (February 2007).

20. Reij, C., Tappan, G., and Belemire, A., Changing Land Management Practices and Vegetation on the Central Plateau of Burkina Faso (1968–2002). *Journal of Arid Environments* 63, 642–659 (2005).

21. Tappan, G., 2009. See also WRI, 2008.

22. Reij, C., and Smaling, E. M. A., Analyzing Successes in Agriculture and Land Management in Sub-Saharan Africa: Is Macro-Level Gloom Obscuring Positive Micro-Level Change? *Land Use Policy* 25, 410–420 (2008).

23. Paeth, H., Born, K., Girmes, R., Podzun, R., and Jacob, D., Regional Climate

Change in Tropical and Northern Africa due to Greenhouse Forcing and Land Use Changes. *Journal of Climate* 22, 114–132 (2009).

24. Butt, T. A., McCarl, B. A., Angerer, J., Dyke, P. T., and Stuth, J. W., The Economic Food Security Implications of Climate Change in Mali. *Climatic Change* 68, 355–378 (2005).

25. Sullivan, G. R. et al., National Security and the Threat of Climate Change, edited by S. Goodman (CNA, Alexandria, VA, 2007), pp. 63.

26. Chakravarty, S. et al., Sharing Global CO2 Emission Reductions Among One Billion High Emitters. *PNAS* 106 (43), 11884–11888 (2009).

27. Bierbaum, R., and Fay, M., 2010.

28. Ki Moon, B., A Climate Culprit in Darfur, in *The Washington Post* (June 16, 2007).

29. Mabey, N. Delivering Climate Security: International World Responses to a Climate Changed World, in *Whitehall Paper*, edited by RUSI (Routledge Journals, Oxford, U.K., 2008), Vol. 69.

30. Burke, M. B. B., Miguel, E., Satynath, S., Dykema, J. A., and Lobell, D., Warming Increases the Risk of Civil War in Africa. *PNAS* 106 (49), 20670–20674 (2009).

## 6. THE GREAT BARRIER REEF, AUSTRALIA

1. Veron, J. E. N., *A Reef in Time: The Great Barrier Reef from Beginning to End* (Belknap Press of Harvard University Press, Cambridge, MA, 2008).

2. Hoegh-Guldberg, O. et al., Coral Reefs under Rapid Climate Change and Ocean Acidification. *Science* 318 (5857), 1737–1742 (2007). See also Kleypas, J. A., Buddemeier, R. W., and Gattuso, J. P., The Future of Coral Reefs in an Age of Global Change. *International Journal of Earth Sciences* 90, 426 (2001). See also Kleypas, J. A., and Eakin, C. M., Scientists' Perceptions of Threats to Coral Reefs: Results of a Survey of Coral Reef Researchers. *Bulletin of Marine Science* 80 (No. 2), 419–436 (2007).

3. Carpenter, K. E., et al., One-Third of Reef-Building Corals Face Elevated Extinction Risk from Climate Change and Local Impacts. *Science* 321 (5888), 560–563 (2008).

4. Scott C. Doney, V. J. F., Richard A. Feely, Joan A. Kleypas, Ocean Acidification: The Other $CO_2$ Problem. *Annual Review of Marine Science* Vol. 1, 169–192 (2009).

5. De'ath, G., Lough, J. M., and Fabricius, K. E., Declining Coral Calcification on the Great Barrier Reef. *Science* 323 (5910), 116–119 (2009).

6. Caldeira, K., and Wickett, M. E., Ocean Model Predictions of Chemistry Changes from Carbon Dioxide Emissions to the Atmosphere and Ocean. *Journal of Geophysical Research Oceans* 110, C09S04 (2005). See also Caldeira, K., and Wickett, M. E., Anthropogenic Carbon and Ocean pH. *Nature* 425, 365 (2003).

7. Lough, J. M., 10th Anniversary Review: A Changing Climate for Coral Reefs. *Journal of Environmental Monitoring* 10 (1), 21–29 (2008).

8. Marshall, P., and Schuttenberg, H., *A Reef Manager's Guide to Coral Bleaching* (Australian Government Great Barrier Reef Marine Park Authority, Townsville, Australia, 2006), p. 176.

9. NOAA, Available at http://coralreefwatch.noaa.gov/ (2009).

10. Henson, B., Heat, Fire, and Fear in Australia, in *UCAR Magazine* (UCAR, 2009).

11. Interim Report, edited by State Government of Victoria (Victorian Bushfires Royal Commission, Victoria, Australia 2009).

## 7. CENTRAL VALLEY, CALIFORNIA

1. Lund, J., et al., *Envisioning Futures for the Sacramento-San Jaoquin Delta* (Public Policy Institute of California, San Francisco, CA, 2007).
2. Vicuna, S., Hanemann, M., and Dale, L., 2006.
3. Lund, J., et al., *Comparing Futures for the Sacramento-San Joaquin Delta* (Public Policy Institute of California, San Francisco, CA, 2008), p. 147.
4. Mastrandrea, M. D., Tebaldi, C., Snyder, C. P., and Schneider, S. H., 2009.
5. Cobb, K., Charles, C. D., Cheng, H., and Edwards, R. L., El Nino/Southern Oscillation and Tropical Pacific Climate during the Last Millennium. *Nature* 424, 271–276 (2003).
6. Guilyardi, E., et al., Understanding El Niño in Ocean-Atmosphere General Circulation Models: Progress and Challenges. *BAMS* 3, 325–340 (2009).
7. Luers, A. L., Cayan, D. R., Franco, G., Hanemann, M., and Croes, B., 2006.
8. Rajagopalan, B., et al., Water Supply Risk on the Colorado River: Can Management Mitigate? *Water Resources Research* 45 (8) (2009).

## 8. THE ARCTIC, PART ONE: INUIT NUNAAT

1. *ITK 5000 Years of Inuit History and Heritage*, edited by Inuit Tapiriit Kanatami (Inuit Tapiriit Kanatami, Ottawa, Canada, 2009).
2. Diamond, J., *Collapse: How Societies Choose to Fail or Succeed* (Penguin Group, New York, 2005).
3. Hall, C. F., *Life with the Esquimaux* (Sampson Low, Son and Marston, London, 1864).
4. Ford, J., et al., Reducing Vulnerability to Climate Change in the Arctic: The Case of Nunavut, Canada. *Arctic* 60 (2), 150–166 (2007).
5. Fox, S., "These Are Things That Are Really Happening": Inuit Perspectives on the Evidence and Impacts of Climate Change in Nunavut, in *The Earth Is Faster Now: Indigenous Observations of Arctic Environmental Change*, edited by I. Krupnik and D. Jolly (Arctic Research Consortium of the United States, Fairbanks, Alaska, 2002), pp. 12–53.
6. Laidler, G., Ford, J., Gough, W. A., and Ikummaq, T., Assessing Inuit Vulnerability to Sea Ice Change in Igloolik, Nunavut. *Climatic Change* (in press, 2009).
7. INAC, The Food Mail Program, Available at http://www.ainc-inac.gc.ca/nth/fon/fm/index-eng.asp (2009).
8. Orlove, B., Chiang, J. C. H., and Cane, M. A., Ethnoclimatology in the Andes: A Cross-Disciplinary Study Uncovers a Scientific Basis for the Scheme Andean Potato Farmers Traditionally Use to Predict the Coming Rains. *American Scientist* 90 (5), 428–435 (2002). See also Orlove, B. S., Chiang, J. C. H., and Cane, M. A., Forecasting Andean Rainfall and Crop Yield from the Influence of El Niño on Pleiades Visibility. *Nature* 403, 68–71 (2000). See also Ford, J. D., Pearce, T., Gilligan, J., Smit, B., and Oakes, J., Climate Change and Hazards Associated with Ice Use in Northern Canada. *Arctic, Antarctic, and Alpine Research* 40 (4), 647–659 (2008).

9. Pearce, T., et al., Inuit Vulnerability and Adaptive Capacity to Climate Change in Ulukhaktok, Northwest Territories, Canada. *Polar Record* 45 (1–21) (2009).

10. Condon, R., Ogina, J., and Elders, H., *The Northern Copper Inuit: A History* (University of Toronto Press, Toronto, 1996).

11. NWMB, 2004.

12. Ainley, D. G., Tynan, C. T., and Stirling, I., Sea Ice: A Critical Habitat for Polar Marine Animals and Birds, in *Sea Ice: An Introduction to Its Physics, Chemistry, Biology, and Geology*, edited by D. N. Thomas and G. S. Dieckmann (Blackwell Science, Oxford, U.K. 2003), pp. 240–266.

13. Ferguson, S. H., Climate Change and Ringed Seal (Phoca hispida) Recruitment in Western Hudson Bay. *Marine Mammal Science* 21 (1), 121–135 (2005).

14. Krupnik, I., and Jolly, D., eds., *The Earth Is Faster Now: Indigenous Observations of Arctic Environmental Change* (Arctic Research Consortium of the United States, Fairbanks, Alaska, 2002).

15. Manabe, S., and Stouffer, R. J., Sensitivity of a Global Climate Model to an Increase of CO2 in the Atmosphere. *Journal of Geophysical Research* 85 (C10), 5529–5554 (1980).

16. Holland, M. M., and Bitz, C. M., Polar Amplification of Climate Change in Coupled Models. *Climate Dynamics* 21 (221–232) (2003).

17. Serreze, M. C., Barrett, A. P., Stroeve, J. C., Kindig, D. N., and Holland, M. M., The Emergence of Surface-Based Arctic Amplification. *The Cryosphere* 3, 11–19 (2009).

18. NSIDC, Arctic Sea Ice News and Analysis (2008).

19. Serreze, M., and Stroeve, J. C., Standing on the Brink. *Nature Reports Climate Change* (2008).

20. Gearheard, S., 2009.

21. Stroeve, J. C., Holland, M. M., Meier, W., Scambos, T., and Serreze, M., Arctic Sea Ice Decline: Faster than Forecast. *Geophysical Research Letters* 34 (L09501) (2007).

22. Maslanik, J. A., et al., A Younger, Thinner Arctic Ice Cover: Increased Potential for Rapid, Extensive Sea-Ice Loss. *Geophysical Research Letters* 34 (L24501) (2007).

23. Zhang, X., and Walsh, J., *Journal of Climate* 19 (2006).

24. Holland, M. M., Bitz, C. M., and Tremblay, B., Future Abrupt Reductions in the Summer Arctic Sea Ice. *Geophysical Research Letters* 33 (L23503) (2006).

**9. THE ARCTIC, PART TWO: GREENLAND**

1. Diamond, J., *Collapse: How Societies Choose to Fail or Succeed* (Penguin Group, New York, 2005).

2. Gautier, D. L., et al., Assessment of Undiscovered Oil and Gas in the Arctic. *Science* 324 (5931), 1175–1179 (2009).

3. Lenton, T. M., et al., Tipping Elements in the Earth's Climate System. *105* (6), 1786–1793 (2008).

4. Funk, M., Greenland Rising, in *Outside* (2009).

5. Saabye, H. E., Brudstykker sd en dagbog holden i Groenland i aarene 1770–1778, in *Medd. Gronland*, edited by H. Oesterman (1942).

6. van Loon, H., and Rogers, J. C., The See-Saw in Winter Temperatures Between Greenland and Northern Europe. Part I. *Monthly Weather Review* 104 (1978).

7. Hurrell, J. W., and Deser, C., North Atlantic Climate Variability: The Role of the North Atlantic Oscillation. *Journal of Marine Systems* 78 (1), 28–41 (2009).

8. Johannessen, O. M., Khvorostovsky, K., Miles, M. W., and Bobylev, L. P., Recent Ice-Sheet Growth in the Interior of Greenland. *Science Express* (2005).

9. Zwally, H. J., et al., Mass Changes of the Greenland and Antarctic Ice Sheets and Shelves and Contributions to Sea-Level Rise: 1992–2002. *Journal of Glaciology* 51 (175), 509–527 (2005).

10. Hanna, E., et al., Runoff and Mass Balance of the Greenland Ice Sheet: 1958–2003. *Journal of Geophysical Research* 110 (D13108) (2005).

11. Pritchard, H. D., Arthern, R. J., Vaughan, D. J., and Edwards, L. A., Extensive Dynamic Thinning on the Margins of the Greenland and Antarctic Ice Sheets. *Nature* 461 (7263) (2009).

12. Bindschadler, R. *Science* 311, 1720–1721 (2006).

13. Krabill, W., et al., Greenland Ice Sheet: High-Elevation Balance and Peripheral Thinning. *Science* 289, 428–430 (2000).

14. Rignot, E., and Kanagaratnam, P., Changes in the Velocity Structure of the Greenland Ice Sheet. *Science* 311 (5763), 986–990 (2006).

15. Ekstrom, G., Nettles, M., and Tasi, V. C., Seasonality and Increasing Frequency of Greenland Glacial Earthquakes. *Science* 311, 1756–1758 (2006).

16. Joughin, I., Abdalati, W., and Fahnestock, M., Large Fluctuations in Speed on Greenland's Jakobshavn Isbræ Glacier. *Nature* 432, 608–610 (2004).

17. Steffensen, J. P., et al., High-Resolution Greenland Ice Core Data Show Abrupt Climate Change Happens in Few Years. *Science* 321 (5889), 680–684 (2008).

18. Fluckiger, J., Climate Change: Did You Say "Fast"? *Science* 321 (5889), 650–651 (2008).

19. Stouffer, R. J., et al., GFDL's CM2 Global Coupled Climate Models. Part IV: Idealized Climate Response. *Journal of Climate* 19 (5), 723–740 (2006). See also Fluckiger, J., Knutti, R., White, J. W. C., and Renssen, H., Modeled Seasonality of Glacial Abrupt Climate Events. *Climate Dynamics* 31, 633–645 (2008).

20. Kukla, G. J., et al., Last Interglacial Climates. *Quaternary Research* 58, 2–13 (2002). See also Oppo, D. W., McManus, J. F., and Cullen, J. L., Evolution and Demise of the Last Interglacial Warmth in the Subpolar North Atlantic. *Quaternary Science Reviews* 25, 3268–3277 (2006).

21. Huybrechts, P., and De Wolde, J., The Dynamic Response of the Greenland and Antarctic Ice Sheets to Multiple-Century Climatic Warming. *Journal of Climate* 12, 2169–2188 (1999).

22. Stroeve, J., Holland, M. M., Meier, W., Scambos, T., and Serreze, M., Arctic Sea Ice Decline: Faster than Forecast. *Geophysical Research Letters* 34, L09501 (2007).

23. Hansen, J. E., A Slippery Slope: How Much Global Warming Constitutes "Dangerous Anthropogenic Interference"? An Editorial Essay. *Climatic Change* 68, 269–279 (2005).

24. Otto-Bliesner, B. L., et al., Simulating Arctic Climate Warmth and Icefield Retreat in the Last Interglaciation. *Science* 311, 1751–1753 (2006).

25. Portions of Arctic Coastline Eroding, No End in Sight, Says New Study (*Science-Daily*, December 17, 2009).

26. Kirby, R. R., and Beaugrand, G., Trophic Amplification of Climate Warming. *Proceedings of the Royal Society B* 276 (1676) (2009).

27. *ICON Carbon Dioxide Capture and Storage Initiative: A Canadian Clean Energy Opportunity*, edited by ICON (2009), p. 16.

28. *DOE Methane Hydrate: Future Energy Within Our Grasp*, edited by Office of Fossil Energy (U.S. Department of Energy, Washington, DC, 2007).

29. Walter, K. M., Zimov, S. A., Chanton, J. P., Verbyla, D., and Chapin III, F. S., Methane Bubbling from Siberian Thaw Lakes as a Positive Feedback to Climate Warming. *Nature* 443, 71–75 (2006).

30. Kennedy, M., Mrofka, D., and von der Borch, C., Snowball Earth Termination by Destabilization of Equatorial Permafrost Methane Clathrate. *Nature* 453, 642–645 (2008).

31. Walter Anthony, K., Methane: A Menace Surfaces, in *Scientific American* (2009), Vol. 301, pp. 68–75.

## 10. DHAKA, BANGLADESH

1. Webster, P. J., Moore, A., Loschnigg, J., and Leban, M., Coupled Ocean-Atmosphere Dynamics in the Indian Ocean during 1997–98. *Nature* 401, 356–360 (1999).

2. Del Ninno, C., Dorosh, P. A., Smith, L. C., and Roy, D., 2001.

3. *NAPA National Adaptation Programme of Action* (NAPA) (Ministry of Environment and Forest Government of the People's Republic of Bangladesh, 2005), p. 46.

4. Cruz, R. V., et al., Asia. *Climate Change 2007: Impacts, Adaptation and Vulnerability. Contribution of Working Group II to the Fourth Assessment Report of the Intergovernmental Panel on Climate Change*, edited by M. L. Parry et al. (Cambridge, U.K., 2007), pp. 469–506.

5. Webster, P. J., et al., A Three-Tier Overlapping Prediction Scheme: Tools for Strategic and Tactical Decisions in the Developing World, in *Predictability of Weather and Climate*, edited by T. N. Palmer (Cambridge University Press, 2006), pp. 645–673.

6. *UNDP World Population Prospects: The 2008 Revision* (United Nations Population Division, 2008).

7. *MEF Bangladesh Climate Change Strategy and Action Plan 2008* (Ministry of Environment and Forests Government of the People's Republic of Bangladesh, Dhaka, Bangladesh, 2008), p. 68.

8. Christensen, J. H., et al., *2007: Regional Climate Projections in Climate Change 2007: The Physical Science Basis. Contribution of Working Group I to the Fourth Assessment Report of the Intergovernmental Panel on Climate Change*, edited by S. Solomon et al. (Cambridge, U.K., 2007).

9. Agrawala, S., Ota, T., Ahmed, A. U., Smith, J., and van Aalst, M., 2003.

10. Colette, A. *Case Studies on Climate Change and World Heritage* (UNESCO, Paris, 2007), p. 79.

11. Friedman, L., Bangladesh: Where the Climate Exodus Begins, in *Greenwire* (2009).

12. Mool, P. K., et al., *Inventory of Glaciers, Glacial Lakes, and Glacial Lake Outburst Floods: Monitoring and Early Warning Systems in the Hindu Kush-Himalayan Region*, edited by ICIMOD (ICIMOD, Bhutan, 2001), p. 254.

13. Nayar, A., When the Ice Melts. *Nature* 461, 1042–1046 (2009).
14. Rodell, M., Velicogna, I., and Famiglietti, J.S., Satellite-Based Estimates of Groundwater Depletion in India. *Nature* 460, 999–1002 (2009).
15. Cruz, R. V., et al., Asia. *Climate Change 2007: Impacts, Adaptation and Vulnerability. Contribution of Working Group II to the Fourth Assessment Report of the Intergovernmental Panel on Climate Change*, edited by M.L. Parry et al. (Cambridge, U.K., 2007), pp. 469–506.
16. Lobell, D. B., et al., Prioritizing Climate Change Adaptation Needs for Food Security in 2030. *Science* 319, 607–610 (2008).

**11. NEW YORK, NEW YORK**

1. Eddy, D. Y2K Discussion List, in Y2K *Discussion List*, edited by Peter de Jager's (1995).
2. Rosenzweig, C., and Solecki, W., eds., *Metro East Coast, Report for the U.S. Global Change Research Program* (Columbia Earth Institute, New York, 2001).
3. DEP, 2008.
4. Rosenzweig, C., et al., Climate Risk Information, in *New York City Panel on Climate Change* (New York, 2009).
5. Rosenzweig, C., and Solecki, W., Climate Change and a Global City: The Potential Consequences of Climate Variability and Change, in *Metro East Coast Assessment* (Columbia Earth Institute, New York, 2001), p. 207.
6. *PlaNYC PlaNYC: A Greener, Greater, New York* (Mayor's Office of Long-Term Planning and Sustainability, New York, NY, 2007).
7. Rosenzweig, C., et al., Managing Climate Change Risks in New York City's Water System: Assessment and Adaptation Planning. *Mitigation and Adaptation Strategies for Global Change* 12 (8), 1391–1409 (2007).
8. *PlaNYC PlaNYC Progress Report 2009* (Mayor's Office of Long-Term Planning and Sustainability, New York, NY, 2009), p. 43.
9. Gray, W., *Landfall Probability Table in United States Landfalling Hurricane Probability Project*, edited by Tropical Meteorology Research Project (Fort Collins, CO, 2009).
10. Chan, S., Why the Subways Flood, in *The New York Times* (August 8, 2007).
11. USGS National Assessment of Coastal Vulnerability to Future Sea-level Rise, in *USGS Fact Sheet* FS-076–00 (Washington, DC, 2000).
12. Ackerman, F., and Stanton, E. A., *Climate Change and the U.S. Economy: The Costs of Inaction* (Tufts University and Stockholm Environment Institute, Medford, MA, 2008).
13. Yin, J., Schlesinger, M. E., and Stouffer, R. J., Model Projections of Rapid Sea-level Rise on the Northeast Coast of the United States. *Nature Geoscience* 2, 262–266 (2009).
14. Kopp, R. E., Personal Communication (Princeton, NJ, Dec. 2, 2009).

**EPILOGUE**

1. Weiss, H., et al., The Genesis and Collapse of 3rd Millenium North Mesopotamian Civilization. *Science* 261, 994–1004 (1993).
2. Cullen, H. M., and deMenocal, P. B., The Possible Role of Climate in the Collapse

of the Akkadian Empire: Evidence from the Deep Sea. *Geology* 20 (8), 379–382 (2000).

3.  *UNFCCC United Nations Framework Convention on Climate Change* (United Nations Environment Programme/World Meteorological Organization Information Unit on Climate Change [IUCC] on behalf of the Interim Secretariat of the Convention, Switzerland, 2002).

4.  Solomon, S., Garcia, R. R., Rowland, S. F., and Wuebbles, D. J., On the Depletion of Antarctic Ozone. *Nature* 321, 755–758 (1986).

5.  Solomon, S., Plattner, G., and Friedlingstein, P., Irreversible Climate Change Due to Carbon Dioxide Emissions. *PNAS* 106 (6), 1704–1709 (2009).

6.  IPCC Physical Science Basis. Contribution of Working Group I to the Fourth Assessment Report of the Intergovernmental Panel on Climate Change in *Climate Change 2007*, edited by S. Solomon et al. (Cambridge University Press, Cambridge, 2007), p. 996.

7.  Revkin, A. C., Scientist at Work: Susan Solomon; Melding Science and Diplomacy to Run a Global Climate Review, in *The New York Times* (February 2007).

8.  Diamond, J., *Collapse: How Societies Choose to Fail or Succeed* (Penguin Group, New York, 2005).

9.  Frame, D. J., et al., Warming Caused by Cumulative Carbon Emissions Towards the Trillionth Tonne. *Nature* 458, 1163–1166 (2009).

10. Meinshausen, M., et al., Greenhouse-Gas Emission Targets for Limiting Global Warming to 2°C. *Nature* 458, 1158–1162 (2009).

11. *EIA International Energy Outlook 2009* (EIA, 2009), p. 274.

## APPENDIX

1.  Meehl, G. A., Tebaldi, C., Walton, G., Easterling, D., and McDaniel, L., The Relative Increase of Record High Maximum Temperatures Compared to Record Low Minimum Temperatures, in the U.S. *Geophysical Research Letters* (in press).

2.  *CCSP Weather and Climate Extremes in a Changing Climate, Regions of Focus: North America, Hawaii, Caribbean and U.S. Pacific Islands*, edited by T. R. Karl et al. (Department of Commerce, NOAA's National Climatic Data Center, Washington, DC, 2008), p. 164.

3.  Meehl, G. A., and Tebaldi, C., More Intense, More Frequent, and Longer Lasting Heat Waves in the 21st Century. *Science* 305 (5686), 994–997 (2004).

4.  Arnell, N. W., *Global Warming, River Flows and Water Resources* (Wiley, Chichester, U.K., 1996). See also Gleick, P. H., Methods for Evaluating the Regional Hydrologic Effects of Global Climate Changes. *Journal of Hydrology* 88, 97–116 (1986).

5.  Zimmernan, R., and Cusker, M., Institutional Decision-making, in *Climate Change and a Global City: The Potential Consequences of Climate Variability and Change. Metro East Coast*, edited by C. Rosenzweig and W. D. Solecki (Columbia Earth Institute New York, 2001).

6.  MacCracken, M. C., Moore, F., and Topping, J. C., eds., *Sudden and Disruptive Climate Change* (Earthscan, London, 2008). See also MacCracken, M. C., Prospects for Future Climate Change Reasons for Early Action. *Journal of the Air and Waste Management Association* 58, 735–786 (2008).

# ACKNOWLEDGMENTS

Rob Socolow, a physics professor at Princeton University who has thought long and hard about how to solve the problem of global warming, adapted the words of Winston Churchill to describe the challenge the world is currently facing. "Never has the work of so few scientists led to so much being asked of so many." I'm struck by these words, because it isn't very often that a scientific finding so thoroughly challenges the foundation on which modern society rests. The discovery of global warming has certainly resulted in an enormous request on behalf of the scientific community to change the way we do business. And although it is not a problem that can be solved overnight, I firmly believe we will get there.

I would like to sincerely thank my colleagues at Climate Central: Berrien Moore, Joanne Graziano, Ben Strauss, Michael Lemonick, Eric Larson, Claudia Tebaldi, Remik Ziemlinski, Nicole Heller, Phil Duffy, Jessica Harrop, Paul Ferlita, Iveta Weinberg, and Andrew Freedman. Science journalism has been hard hit in the past year, and I applaud the hard work of my fellow scientists and journalists engaged in the task of communicating climate science and technology. Climate Central was established to provide unbiased information about climate science in order to help people better understand what the future may hold if we continue to have an infrastructure based only on fossil fuels. Without the guidance and support of Climate Central's founding board members, this organization could not have moved from concept to reality. For engaging in this vision of a new model of science journalism, I would like to thank Wendy Schmidt; Joe Sciortino and the Schmidt Family Foundation; Steve Pacala at Princeton University; Gus Speth at Yale University; and

Jane Lubchenco, who has long since retired from the board and become chief administrator at NOAA. I would also like to thank Climate Central's original board members: Richard Somerville at the Scripps Institution of Oceanography; Michael Oppenheimer at Princeton University; Sally Benson, Pamela Matson, and Jon Krosnick at Stanford University; Skip Lupia at the University of Michigan; Mary Evelyn Tucker at Yale University; and John Holdren, who has since retired from the board to serve as director of the Office of Science and Technology Policy.

My most heartfelt thanks to the scientists who shared their lives and their time with me: Stephen Schneider at Stanford University; Susan Solomon at NOAA's Earth System Research Laboratory; Joanie Kleypas at the National Center for Atmospheric Research; Cynthia Rosenzweig at NASA's Goddard Institute for Space Studies; Steve Hammer at Columbia University; Ellen Hanak at the Public Policy Institute of California; Jay Lund at the University of California, Davis; Peter Webster at the Georgia Institute of Technology; Omar Rahman at Independent University in Dhaka, Bangladesh; Tristan Pearce at the University of Guelph; Shari Gearheard of the National Snow and Ice Data Center at the University of Colorado at Boulder; Alessandra Giannini at the International Research Institute for Climate and Society; Chris Reij at Vrije Universiteit; Isaac Held at NOAA's Geophysical Fluid Dynamics Laboratory (GFDL); J. P. Steffensen and Dorthe Dahl-Jensen at the University of Copenhagen; Scott Luthcke at the NASA Goddard Space Flight Center; and John Chiang at the University of California, Berkeley.

I also need to thank Mike MacCracken of the Climate Institute for carefully reading through this manuscript; his knowledge of climate science is truly encyclopedic. In addition, I'm grateful to Jerry Meehl and Peter Gent at NCAR, Ron Stouffer at NOAA's GFDL, Peter deMenocal at the Lamont-Doherty Earth Observatory of Columbia University, Vasilii Petrenko at the University of Colorado, Michael Bender at Princeton University, and Phil Arkin at the University of Maryland—all very busy scientists who were willing to lend their expertise and feedback.

I'm so very grateful to Marla Hoppenfeld and everyone involved in this project at HarperCollins Publishers. And I am especially grateful to Matt Harper, my editor—without his continued support and guidance, this book would have never become a reality. My thanks also go out to Lisa Sharkey for helping to shepherd this project through a tough economic landscape.

I must also thank my wonderful and supportive family: William and Heidi Cullen; Rosemary Wenke and Jim Honrine; and Stephen, Donna, Matt, and Keith Cullen. If it weren't for my mom and dad, who came to my rescue on the weekends to help around the house (and bake apple pie!), this project would have been a whole lot harder if not altogether impossible.

And finally, the project would have been truly impossible without the love, support, and friendship of my husband, Drew. He is my rock and my hero. And to our pooches, Homer and Emma, here's to finally taking some long walks again!

Heidi Cullen
Princeton, New Jersey, January 2010

# INDEX

extreme weather, 50–56
  Great Barrier Reef, 98–99
  New York City, 233–34
Exxon Mobil, 177

famine, 78–79, 81–87, 204–5
farmer-managed natural regeneration
  (FMNR), 77–78
feedback, 21–22
*Finding Nemo* (movie), 103
Flood Forecasting Warning Center
  (FFWC), 202–3
floods (flooding), 54–55, 121–22,
  221
  Bangladesh, 197–204, 208–13,
    219–25
  Central Valley, 121–22, 124–30
  New Orleans, 121–22, 133–34,
    249
  New York City, 236–44, 255–56,
    258
  Red River, 3–5, 10
Food Mail, 157
forcings, 21–22, 38, 44–49. *See also*
  solar radiation; volcanoes
forecasting (forecasts). *See also* climate
  forecasting; weather forecasting
  linking weather and climate, 8–11
forecast mode, 41–42, 53
Fourier, Joseph, 18–19, 20, 21
Franklin, Benjamin, 45
Frederick IV of Denmark, 176
free radicals, 95–96
Friends Ambulance Unit, 33

Ganges River, 201–2, 209, 211, 213
Gangotri Glacier, 209
gas chromatographs, 27–28
Gearheard, Jake, 155–57, 161
Gearheard, Shari, 149–58, 161–62,
  165, 168–72
genetic engineering, 109–10

Geophysical Fluid Dynamics
  Laboratory (GFDL), 72–74, 80
Georgia Institute of Technology,
  197–98
Gere, Richard, 115
Giannini, Alessandra, 68–72, 74–75
glacial lake outburst flood (GLOF),
  220
glaciation, 16–18
glacier melt, 179–88, 209, 211,
  220–21
  Arctic forty-year forecast, 188–95
global warming, xv–xviii
  Arrhenius and, 25–26, 30
  Bangladesh, 203, 211–14
  Central Valley, 121–37
  climate model and, 39–40, 42–45
  extreme weather, 50–56
  Great Barrier Reef, 89–90, 93–104
  Greenland, 177–88
  Inuit Nunaat, 154–66
  New York City, 229–41
  ocean surface temperature and,
    72–73
  public attitude about, 5–8, 10–11
  Sahel region, 63, 72–74
  stock market compared with, 7
Goddard Institute for Space Studies
  (GISS), 41–42, 230–31, 254
Gold Rush, 119–20
GPS maps, and Inuit, 169–71
Gravity Recovery and Climate
  Experiment (GRACE), 180–81,
  184
Great Barrier Reef, Australia, 89–114
  adaptation strategies, 104–10
  coral bleaching, 95–100
  forty-year forecast, 110–14
  global warming and, 89–90,
    93–104
  history of, 91–93
  map, 88